The Blue Parka Man

The Blue Parka Man

ALASKAN GOLD RUSH BANDIT

H. C. Landru

Illustrated with photographs

DODD, MEAD & COMPANY • New York

Printed in the United States of America

1 2 3 4 5 6 7 8 9 10

Library of Congress Cataloging in Publication Data

Landru, H C
  The blue parka man.

    1. Hendrickson, Charles.   2. Outlaws—Alaska—
Fairbanks region—Biography.   3. Fairbanks region—
History.   4. Fairbanks region—Biography.   I. Title.
F914.F16H465   979.8'603'0924 [B]   79-25575
ISBN 0-396-07821-4

To my son who, as a boy,
was an eager seeker
among the eerie crags of Cleary Summit
for the treasure of the Blue Parka Man

LIST OF ILLUSTRATIONS

CONTENTS

AUTHOR'S PREFACE

I was educated as a historian with A.B. and M.A. degrees from the University of Oregon separated by a year's teaching experience at the Alaska Agriculture College and School of Mines, predecessor of the University of Alaska. Then followed three years as Sanders Fellow at The George Washington University, Washington, D.C., and another three years as historian and superintendent with the National Park Service in the national historical parks before the lure of the North called my wife and me back to Alaska. We prospected for gold on the high tributaries of the middle Kuskokwim until the outbreak of World War II when we moved to Fairbanks. There I pursued a twenty-eight-year civilian career with the U.S. Air Force and Army as budget officer while we homesteaded, raced dogs, and my wife continued her career as an author.

When I first began collecting material for *The Blue Parka Man*, a score of the men who witnessed or participated in the events described in the book were still alive. Many of the incidents and conversations were based on stories told to me by these old-timers. Among them were Tom Gibson, the caribou hunter; dapper little Billy Gorbracht, the dance hall musician; Al Young who worked in one of the mines with Hendrickson; Frank P. Young who knew Deputy U.S. Marshal George Dreibelbis; Deputy U.S. Marshal Frank Wiseman; and Fred Parker of the Caroll–Parker sawmill firm, who was the first mayor of Fairbanks and, incidentally, my father-in-law.

INTRODUCTION

Wilderness had a very different meaning to the men of stampede Alaska than we generally attribute to it today. To them the wilderness was not a friend. It was an adversary, an antagonist, to be respected, often feared, and only subdued when a camp gave promise of enough permanence to justify the incredible cost.

The numbing winter temperatures, sometimes dropping to seventy below zero, stealthily drained a man's strength; he could freeze without ever being conscious of it. And there were many other dangers. Deadly overflows hidden under a blanket of snow would continue to flow above the glaciated stream beds; the unwary traveler who blundered into this trap often perished before he could get a life-giving fire started. The absence of distinguishable landmarks in a land of stunted spruce and birch covered with an unvarying mantle of snow was yet another threat.

Even when this frozen land was transformed as though by a miracle in a short month's time to a virtual carpet of blooming Russian honeysuckle, salmon berries, wild currants, and cranberries, man could not relax and experience the new creation. The twenty hours of sunlight each day thawed th lakes and loosed the rivers, bringing only different hardships. Mosquitoes rose in swarms from the thawing muskeg; bears and moose, to say nothing of man, have been driven mad by their insistent attack. The valley bottoms, at least snow-packed to a level grade in winter became nearly impassable mires. Bad enough that they sucked at every footstep, but the muskeg was covered with hummocks of decayed mosses or long-dead stumps so closely and irregularly spaced that a man would slip with every step. Traveling the valleys was merciless going. Nor was summer what one might expect—a reasonable transformation from the winter's frigid cold. With near-ceaseless daylight, temperatures frequently soared to the low and mid-nineties. In this sun-drenched luxury the land quivered with life, reaching almost subtropical growth. But man's thermostat became confused; he merely suffered.

This was the wilderness from which Fairbanks was hewn. One can imagine no more unlikely location for a town. A pair of calipers and considerable higher mathematics indicates that Fairbanks is the exact geographical center of Alaska. It is situated on the Chena Slough, a river in any other land except here, where rivers and mountains are measured in terms of the length and breadth of the Yukon and the soaring height of McKinley. Ten miles below this townsite, the Chena joins the Tanana River, the largest southern tributary of the Yukon. About 120 miles to the north as the raven flies lies the Arctic Circle.

At this point the Tanana Valley is fifty miles wide. It stretches southward to the rugged skyline of the Alaska Range dominated by Mount McKinley, a massive barrier that seals off interior Alaska from all moderating effects of the coast climate. The Chena Slough, a merging of waters from a meandering channel of the Tanana River and the Chena River, clings to the northern extremity of this wide valley, five to twenty miles from the summits of low rolling hills. These hills, seldom exceeding 2,000 feet elevation, are the deeply eroded remains of an old mountain system that may have rivaled the Alaska Range in early geologic time. Weathering over countless millennia created a golden but well-guarded treasure, the last of the great bonanzas of the North.

Gold is more than a metal. It is the supreme catalyst, undergoing no change but radically altering whatever it touches. It is at once a creator and a destroyer. It is also a disease, and a stampede is its ultimate expression. Word of a gold strike spreads on the wind. The prospectors had a word for it: the mukluk telegraph. To the men of those days, stampeding was a way of life, with every ear tuned to the word *gold*. Trail packs were in constant readiness and, at the first rumor of a strike, whole communities would be deserted in a mad scramble to reach the latest discovery. The stampede was no respecter of class or occupation. Drift miners, woodcutters, trappers, lawyers, bankers, merchants, doctors, dentists, clergymen, pimps and prostitutes, gamblers, saloonkeepers, cooks and waiters—every stratum of society was represented. They all traveled light and they traveled fast. By foot, dog team, horse, boat, mule they came, debtor and creditor, everything forgotten in the rush to be first in the quest for riches.

In June 1905 the *Fairbanks News* reported that approximately six thousand mining claims located on some six hundred creeks had been staked and recorded in the Fairbanks district. Gold washed from the rich creek gravels rose from $40,000 in the first development year of 1903 to $600,000 in 1904, then to $6,000,000 in 1905, and an incredible $9,000,000 in 1906.

As the creeks boomed, the town mushroomed. Transportation was its most critical need. In the first two seasons, the Circle City overland route did well enough, with its roadhouses spaced at twenty-mile intervals to care for the needs of both summer and winter travelers. But as the camp grew, this route became inadequate.

The new routes of transportation were dictated by the seasons. In the ice-free summer months, a fast new fleet of sternwheel steamers, built to draw less water than most of the Yukon boats, appeared on the Tanana and the Chena. These shallower draft boats connected with the Yukon steamers at Tanana and at the Army post of Fort Gibbon, where the Tanana and Yukon rivers joined, some 235 river miles below Fairbanks. This gave access to the Outside via the great broad avenue of the Yukon, with connections at both St. Michael at the river's mouth and Dawson near its upper reaches.

The summers were short, however, and a secondary winter route of transportation had to be developed to supply the camp during the remaining seven months of the year. This was soon established with the building of the Valdez Trail. *Building,* and even the word *trail,* is something of a misnomer in this case, for there was no roadbed and no grading. The winter snow provided the only foundation. That this trail should even have been conceived, much less realized, has to rate as a minor miracle. From the ice-free port of Valdez, the trail threaded Keystone Canyon on the Lowe River. In eighteen miles it climbed 3,000 feet through the Chugach Range to Thompson Pass, whose craggy summit was swept bare by the unrelenting winds in an otherwise boundless world of snow. From Thompson Pass, skirting the Copper River, the trail descended to Paxon and Summit lakes before rising abruptly again to 3,000-foot Isabel Pass, which gave access through the Alaska Range to the Big Delta and Tanana rivers. After 400 treacherous miles the trail reached Fairbanks, a marvel of man's ingenuity and determination.

At first the Valdez Trail was traveled only by dog teams. It was not long until horse-drawn double enders and stages were in operation on a regular schedule, departing both Valdez and Fairbanks at ten-day intervals.

The stages carried the U.S. mail, passengers, and Wells Fargo express. Over twenty-five roadhouses along the trail were soon providing overnight accommodations and meals for the travelers. Bundled in fur coats and robes and comforted with hot-brick foot warmers, these travelers faced few discomforts except in cases of high winds and the frequent, unpredictable storms. Those with a short grubstake, as well as many of the more adventuresome, walked the trail on foot. Even bicycles were used. Granted, this trail might not have been the most expeditious means of travel, but it did provide Fairbanks during the long and sometimes dreary winter months with a regular access to the Outside, reliable mail service, and adequate passenger transportation.

It was not as isolated a life as one might imagine. The Army telegraph system, already linking the far-flung Alaskan military garrisons, was completed to Fairbanks in July 1903. Thereafter, barring winter storms and summer floods and forest fires, which periodically carried away the land lines, Fairbanks received the world's daily news as speedily as San Francisco or Seattle.

Almost all of the physical ingredients of the town were soon present, but not quite all. Both the town and the mines were screaming for lumber for every conceivable need. With abundant spruce forests at hand, sawmills soon appeared. Four mills were doing a land-office business in 1905. The arrival of the first, the Caroll-Parker sawmill, was a true pioneering odyssey.

The mill was operating at Dawson when news of the Fairbanks strike arrived. The partners decided to head for the new camp. During the winter of 1902–1903 they loaded all of their equipment on horse-drawn sleds and set off down the frozen Yukon River. Temperatures often dropped to sixty and more below. At Fortymile, so named because it was located forty miles below old Fort Reliance, they headed up the gorgelike valley of Fortymile

River, finally reaching the summit that divides the Yukon and Tanana drainage. The high uplands crossed, they began the long, slow descent: 170 miles to the Tanana by way of the Goodpasture River, with sled runners frequently wrapped in chains—rough locked—on steep grades to prevent runaway loss of the sleds.

After reaching the Tanana, the men hurriedly set to work building rafts to be ready for the high water of the spring breakup. In the shallow channels of the Tanana, not an inch of water could be lost. With the ice still running strongly, they cast off for the Chena—sawmill, crew, horses, their now rapidly diminishing food supplies and trail gear almost awash on the ungainly rafts. The river current was their only power, and the partners knew that their sole hope of reaching Fairbanks was finding the still unnamed slough reported to connect with the Chena some twenty miles above the town. They could not understand or be understood by the few Indians they encountered, and it seemed that the rumored slough was as elusive as a mirage. Help came finally from an unexpected and almost disastrous source, one of the turbulent windstorms common to the river that whipped the sand from the river bars in a blinding fury of choking dust. The rafts were headed for the relative safety of the riverbank and there, to the men's unbelieving sight, lay the nearly obscured entrance to the slough. A few days later the six-month-long odyssey ended on the booming Fairbanks waterfront. These hardships were not unparalleled, but the accomplishment typifies the spirit that infused and drove the stampeders.

By the summer of 1905 the sawmills were turning out railroad ties in vast numbers. That meant a railroad, and even this fantasy was realized. On July 17, 1905, Fairbanks celebrated completion of the first fourteen-mile division of the line with a spectacular golden-spike—driving ceremony. The Tanana Mines Railroad, later named the

Tanana Valley Railroad, was a reality. Twice-a-day railway service between Chena at the mouth of the Chena Slough, and Fairbanks was immediately in effect. Construction barely paused for the ceremony, however, before pushing on to the creeks. Eventually the railroad operated forty-six miles of line.

Nobody was keeping an accurate head count in those feverish days, but Fairbanks was generally estimated to have a population of 5,000, with 3,000 more people on the adjoining creeks and in Chena City. Some estimates ranged as high as 10,000 for the camp and creeks, more than double the population of all the remainder of Alaska. Its people were of every nationality—Slavonians, Italians, Hungarians, Scots, Japanese, and of course the Swedes and Norwegians; from nearly every state of the Union; from every conceivable social strata and occupation. All were drawn by a single motivating force: to make their fortune.

Almost from the start, Fairbanks was also a town of homes, families, and children. This was largely due to the nature of its gold deposits. In marked contrast to Dawson, where on many creeks gold was found at the grass roots, and in contrast to Nome, where the beach sands almost glittered with the color of gold, most of the gold on Fairbanks creeks was deeply buried under perpetually frozen muck and gravel. This required a substantial investment of capital for machinery and for labor crews.

In drift mining, shafts were sunk to bedrock, with long, low tunnels radiating along the face of the pay. All the mines worked ten-hour shifts; a day's work for a man in the drift was 145 wheelbarrows of dirt from the face of the cut to the shaft. It was worse than mule's work: ten hours spent in a perpetual crouch, in dimly lit caverns, always in mud and water from the thawing face, the roof of the tunnels often perilously supported by only unthawed pillars of gravel. There was never a pause. It took the con-

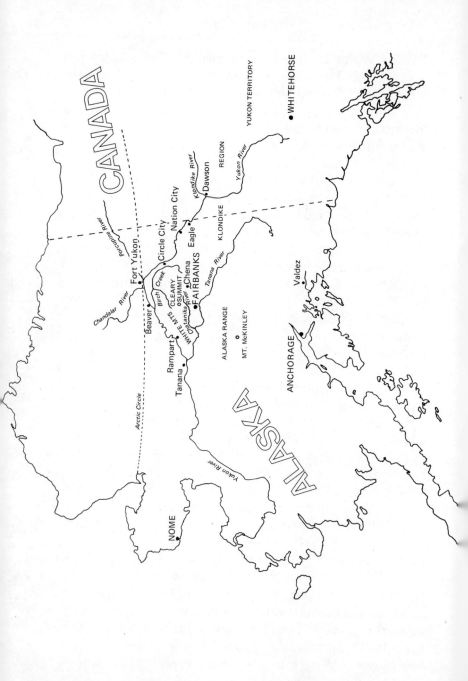

stitution of a gorilla to endure the pace. A man fell asleep immediately after supper, too dead tired to move until it was time for the next shift.

But when a man drew his pay his life was his own. What he sought was excitement, any kind of excitement, so long as it was a complete break from the physical and mental drudgery and monotony of the drift. Among the most elaborate of the entertainment spots—the word *entertainment* had a broad connotation—were the Floradora Dance Hall; the Century Hall, site of the local prize fights; the California, which provided a bowling alley among other attractions for its patrons; and the Tanana Saloon, where proprietor George Butler was always the biggest stakeholder in Fairbanks. These and other establishments might vary in some of the attractions they offered, but all provided every gambling choice the pocket could stand, and more. They offered liquor of all qualities for the most fastidious drinker to the man who just wanted to get drunk fast. If women were not always on the premises, they were available in allied and convenient sporting houses. The entertainment spots offered glamor, bright lights, and excitement—and the fastest means in the world to empty a man's poke. These places never closed; in the twenty-four hour nights of winter, why bother?

The drift workers might have been foremost in their thirst for excitement, but certainly they were not alone. The entire camp craved it. Gambling, drinking, and sporting were only manifestations of this compelling urge. What the camp needed was a rousing, home-grown swell of excitement that would galvanize men's attention, shock them out of their narrow routines and customary pursuits. The old-fashioned western hangings must have had such origins. Fairbanks did not have long to wait. All the ingredients were present for just such a shocker.

1

THE BLUE PARKA MAN

Highway robberies were as foreign to Alaska as palm trees. There were good reasons. Laws were leniently enough enforced so there was no necessity for such desperate chances, better relegated to the vanished days of the western stagecoach. And if Alaskan laws were lenient, the wilderness was not. That was the alternative a man chose. There was no refuge for a hunted man in a town or a creek camp where everyone was well known. Few men willingly chose the solitary challenge of the wilderness; most feared and chose to avoid it.

There was an even better reason to discourage banditry in 1904 and in the early months of 1905. Every man was in hock. The entire camp lived on credit. Even mining crews had been promised their winter pay on bedrock, and any man who could jingle together a couple of quarters was worth a touch.

There was gold enough to satisfy every demand, enough even to have tempted Saint Peter, but it was locked in the winter dumps, small mountains of gravel on every creek ripped from the winter drifts. That early spring everyone was awaiting the return of the sun to transform the frozen creeks first to trickles, then to rushing torrents of water. With the release of the sun's strength, the dumps also thawed. Then men worked feverishly, weeling the golden gravel into waiting sluice boxes where the cascading water trapped the gold in the wooden riffles. Often the gold lay so heavy in the boxes that the miners had to clean up after every shift.

Soon the dust was moving to town to pay bills, to quench the long winter's thirst, to buy a girl or new shirts and boots. The whole camp was on a happy spending binge that spring of 1905. Everyone had money again. But a lot of the gold got sidetracked.

Highway robbery would have been a poorer paying cash business than most the preceding winter. Now it offered an opportunity never encountered before in Alaska—to anyone willing to accept the risk. The first robberies began randomly in April, and by May had settled into a familiar pattern.

The richest creeks—Cleary and Fairbanks—lay just across Cleary Summit, a high divide twenty miles north of Fairbanks. On this bare, craggy summit a man holed out in the rocks commanded a perfect view of all travelers ascending the hog-back trail from Cleary and the eastern slope from Fairbanks Creek, as well as the long, gradual back trail to Fairbanks. Within a few hundred feet on either side of the summit the trail dropped into sheltering timber, but nothing obstructed the view from the high, windy crags.

The road connecting Fairbanks and the placer creeks traversed a lonely, almost uninhabited country. A man could travel the twenty-odd miles and never see another soul except travelers like himself. But as the winter's darkness and penetrating cold gave way to a new world of endless daylight, resinous fragrance, and flaming color, the trail seemed suddenly peopled by a score of highwaymen striking at widely separated points. Robberies became an almost daily occurrence.

From uneasy concern in April, the situation developed to fear in May, then to a state of terror on the part of many of the townsmen by June. Highway robberies occurred with frightening regularity. The victims could never identify their assailants. The highwaymen were always masked, and the victims were generally too shaken by their experience to give anything but the most meager descriptions. Increasingly, however, it became apparent that one highwayman in particular was involved in many of these robberies. He was always identified by a blue drill parka, was invariably described as a powerful, well-built man of remarkable coolness and few words, and was always armed, by those discerning enough to notice, with a .30-.30–caliber Savage. It was not long before he was known throughout the entire region as the Blue Parka Man. His exploits became the favorite subject of the camps' conversations.

Not many firsthand accounts of the Blue Parka Man's activities have persisted through the years, but those that have provide graphic pictures of the events.

On one particularly hot, mosquito-ridden June afternoon, a solitary prospector trudged across the bridge into Fairbanks. The streets were almost empty, with most of the camp's inhabitants doing as little work as possible and sweating out the hours until the evening coolness amid the refreshments offered by the Floradora.

The first man to stop the elderly traveler passed the news. Here was the latest victim of the Blue Parka Man! The saloons emptied and a growing crowd converged on the prospector.

"How much did he lift from you?" was the inevitable first question.

The man looked puzzled. "He bought me a drink," he said finally.

There were muttered exclamations from the crowd and

a laugh or two. The fellow obviously had a touch of the sun, or was bush loco.

Sensing the crowd's reaction, the man shouted, "I'm no crazier than the rest of you galoots! I said he bought me a drink and that's what he did!"

"Let him talk," someone yelled, and things quieted down.

"You know where the trail drops into timber from Pedro Dome? Well, he was waiting for me there, standing still as one of the spruces. Nothing moved but his eyes. I never saw him till he spoke.

" 'Hand over what you've got,' he told me.

"I didn't even try telling him I had only ten bucks to my name. Didn't seem any use in it. I just shelled out my poke and emptied my pockets.

" 'Toss the poke and your pack over here,' he ordered.

"About five seconds and he'd fingered through the

Vide Bartlett Collection in the Archives, University of Alaska, Fairbanks

Spring cleanup on Cleary Creek

whole thing. He stood there with my ten bucks in his hand.

" 'That everything you got?'

"I said yes, and that Savage he kept dead-centered on me every minute looked big as a cannon.

"While I watched him, neither of us saying a word, he put the money back in my poke and tossed it back to me. Then he reached in a pocket and flipped me a four-bit piece.

" 'Buy yourself a drink at the next roadhouse,' he said.

"Then he just backed into the timber and that's the last I ever saw of him. But I am sure buying that drink. Thinking about it is all that ever got me into town."

As was to be expected, the merchants took the dimmest view of these activities. The freighter robbed of his gold—cleanup—or a working stiff who had his wages lifted was more shamefaced than furious. The merchant threatened with diminishing profits was an angry and articulate enemy.

Perhaps one of the most unusual accounts of a meeting with the Blue Parka Man was that of Episcopal Bishop Peter Trimble Rowe. He is the only man known to have argued with the determined parka-clad bandit, but then Bishop Rowe was also a determined man.

The encounter took place one blustery, overcast day when Bishop Rowe and six companions were crossing the Cleary Divide. The Blue Parka Man easily got the drop on the entire party. All were ordered to toss their pokes in his direction. They hastily complied, except Bishop Rowe.

"What do you mean robbing a minister of the gospel?" Rowe demanded.

The bandit gazed steadily at the bishop for several seconds in silence. "Give him your pokes," he ordered the others with a wave of his rifle in the bishop's direction. The pokes were placed one by one in the bishop's hands.

"You're Bishop Rowe," the highwayman stated. "You can use that money better than me. You've just made a donation to the church, boys. Glad to meet you, Bishop." The bandit vanished in a swirl of low drifting clouds. No one knew exactly when he left.

History does not record whether Bishop Rowe kept the Blue Parka Man's donation. Knowing the devotion of the bishop to the welfare of his maverick flock, it is a reasonable assumption that he did.

The stories of these encounters were circulated as widely by the miners and townsmen as were the accounts of the less fortunate victims. Sentiments ranged from horror to admiration, but the majority reaction was amusement. A legend was already growing about the blue parka-clad bandit.

The only man known to have discovered the identity of the Blue Parka Man during the period of the robberies was Billy Gorbracht, a small, dapper German musician. And Billy was not talking. He had known better days as a violinist with the Metropolitan Opera, but what led him to forsake the Metropolitan for the North was his secret. Gold eluded him, but his violin and trumpet found ready demand in the Dawson dance halls, and more recently at the Floradora. On this particular occasion he was returning from a trip to Cleary City where he had played for a church-raising benefit.

Billy glanced apprehensively at the thick tangle of spruces growing close to the trail, near the windswept crest of Cleary Summit. Off to his left the bald granite crags gave cover to every manner of predator—wolves, foxes, and, most to be feared, man.

The warming April sun was fast melting the last of the winter snow, but Billy's hands were numb in the too carefully fitted gloves he fancied. Hurriedly he adjusted the cinches on his packhorse, unmindful of the sun-drenched panorama of Fairbanks, Goldstream, and Cleary creeks spread out on the valley floors 2,000 feet below. He was less concerned with the millions in placer gold already spilling into the stampede camp of Fairbanks than with his own two hundred dollars rolled up in a Bull Durham tobacco sack tied in his horse's tail.

He slapped his horse's rump, and turned to find a .30-.30 Savage pointed squarely at his belt buckle.

Billy shuddered. Violins and trumpets were his business, not guns, and this one in particular in the steady hands of the barrel-chested stranger gave him no confidence. There was no second's doubt that this was the Blue Parka Man.

"You still play a good trumpet, Billy." The voice, tinged with a barely discernible accent, sounded friendly, but the rifle never wavered.

Billy's mind somersaulted back through the years and all the faces it had known.

"Right now you might not remember me," the voice continued. "Be sure you never do. I like your music and the Floradora. I aim to keep on enjoying them."

Obviously the Blue Parka Man had made one of his rare mistakes. Momentary carelessness had betrayed him into stopping a man who might recognize him. Now he was bluffing it out. It didn't take much of a bluff so far as Billy was concerned. He could forget easy.

The Savage was cradled in the nook of an arm now, as naturally as a woman carries her baby.

"Get going." When Billy hesitated, he added, "I don't want anything from you, Billy. Just remember it's healthier if you forget I ever lived."

As suddenly as he had appeared he was gone, and Billy was left alone to continue unmolested the remainder of his solitary trek to Fairbanks.

The first public outcry for concerted community action against the highwaymen came on May 18. Some of the leading citizens of the town, including Episcopal Archdeacon Stuck and Attorney and U.S. Commissioner Charles E. Claypool, started to organize a vigilance committee.

At that time the newly appointed U.S. marshal, George G. Perry, had not yet reached Fairbanks. The office was in charge of Chief Deputy E. E. Reynoldson, with George Dreibelbis its motivating force. Dreibelbis was an old-time Montana lawman from the days of the cattle wars, and one of the most capable law officers in Alaska's history. He not only knew how to quell an incipient riot, he knew where to look for its instigators. Dreibelbis paid a visit to the editor of the *Fairbanks Evening News*. He stated in clear terms that any vigilance committee action constituted an illegal assumption of police powers, and that he would hold the members responsible for their acts. That put an effective damper on the whole idea, and the movement died. It is unfortunate that a man of Dreibelbis's mold did not head the marshal's office. If he had, many of the events to follow could have been avoided, or at least greatly mitigated.

Judge Wickersham was also to play an important role in these events. The day after Dreibelbis's visit to the *Evening News*, the Judge penned a three-page letter to Marshal Perry, to await his arrival. It was also addressed to R. V. Harlan, U.S. District Attorney. In his letter Judge Wickersham detailed the highway robberies of the past six weeks, the threat of violence posed by a vigilance committee, the need for an immediate request to the Attorney General for at least three additional deputy marshals, and immediate, secret action to assemble a posse of twenty men under George Dreibelbis to secure the road to the creeks. Judge Wickersham went on to recommend that members of the criminal class be prosecuted for vagrancy and be jailed, that gambling be suppressed, and that prostitutes be forbidden to carry on their trade in the saloons. He urged energetic action, promised the full

support of his office, and then added further advice:
"This seems a fitting and necessary time for the applica-
tion of President Theodore Roosevelt's 'Big Stick.' "

Clara Rust Collection in the Archives, University of Alaska,
Fairbanks

Treasure train with gold dust for First National Bank, Fairbanks.
Second shipment, spring 1906

Never one to delay action when it seemed imperative,
Judge Wickersham initiated a joint telegram from his
office, the marshal's and the district attorney's to the attorney general in Washington, D.C., asking authorization
for additional deputy marshals. This he followed up with
a personal letter outlining the conditions in Fairbanks.

One week later, on May 27, after a long delay, Marshal
Perry arrived. The *Evening News* reported that its editor
greeted the marshal cordially on his arrival, believing
that his appearance would greatly improve conditions. If

such cordiality ever did exist between the marshal's office and the press—including the *Evening News'* competitors, the *Fairbanks Daily* and *Weekly Times*—it rapidly eroded.

One significant improvement, however, quickly followed, inspired by Judge Wickersham's suggestion for a marshal's posse. On May 31 the Alaska Pacific Express Company entered into an arrangement with the three local banks for armed guards to accompany weekly trips of treasure trains to the creeks. These treasure trains consisted of a string of packhorses and mules accompanied on occasion by as many as twenty armed guards and one deputy marshal. The train made the rounds of the creek mines, picking up the miner's cleanups. Under this vigilant protection the gold was safely convoyed to the banks for deposit. No treasure trains were ever attacked.

Marshal Perry also acted quickly on another of Judge Wickersham's suggestions, but in this case a poorly conceived one. All gambling in the city was closed down. As a result, the temper of the town flared. What Marshal Perry needed was friends; instead he provoked an angry wasps' nest. The action was badly timed and a few days later was rescinded.

In the meantime almost daily holdups of individuals on the trail continued. Also daily, the *Evening News* continued to visit the marshal's office to inquire about progress and to request statements. The marshal, harassed by both highwaymen and the *News,* could be excused for not presenting the best press image. And at about this same time a devastating forest fire swept the Eagle district to the east of Fairbanks, burning the pole lines on which all wire communications depended. With the wires down for three weeks and no wire service to supplement local news, there was ample newspaper space to devote to what was rapidly becoming a running feud with the marshal's office.

In the oftentime callous humor of the frontier, the News reported on May 31. "A Fairbanks Jap had $1,000 in gold dust but it was taken from him, probably by some fellow who will put it in circulation." The incident was just one other in the rising toll of daily robberies.

An event that sent smoldering tempers of the town soaring to new highs occurred June 16 with the most daring of all the robberies to date. It took place almost within sight of the Gilmore Roadhouse, a scant ten miles from Fairbanks. Never before had the highwaymen struck this close to the town.

Two freighters, Norman J. Campbell and a man named Heel, left the Gilmore Roadhouse at about seven o'clock that morning. They were walking beside their wagons climbing a hill, resting every few rods. The two wagons kept close together, with Campbell in the lead. At one of the steep pitches, while resting, Heel put his hand upon the rear dashboard of Campbell's wagon. Campbell turned around to say something, and met the gaze of a masked man who already had Heel covered. The robber had come upon the two men so quickly that neither saw him until he was beside them.

"It was very evident that he understood his business," the freighters later recounted. "He was as cool as though passing the time of day. He did not say over a half dozen words, and handled his rifle in that easy manner which showed he was familiar with its use. It was never pointed directly at either of us, but held so that it covered both of us."

From Campbell's description the man was above average size and wore blue overalls with a slouch white hat. His shirt was black and his face was well protected with a black mask. He did not wear a coat.

After turning over their pokes containing $230, the freighters were ordered to drive on and not to look back. They did glance around, but the man had disappeared into the timber. The robbery was a daring one, not only because of its nearness to Fairbanks and close proximity of the roadhouse, but because other parties were just ahead of and behind the wagons. One man was on top of the hill waiting for the wagons to catch up, and two or three outfits were a short distance behind. Although the highwayman was not distinguished by his customary parka, it was evident from matching descriptions that the freighters had been the most recent victims of the Blue Parka Man.

Fairbanks's already taut nerves were stretched even tighter just over a week later, on June 24, when the highwayman struck at the very outskirts of Fairbanks. Four men were held up near the Owl Roadhouse at the foot of Birch Hill, just two miles from the town limits.

That very evening the *News* launched its most vicious attack yet on the beleaguered marshal's office. A scathing editorial stated that Marshal Perry distinctly conveyed the impression that not only was no action being taken, but that no one in the marshal's office had any knowledge of holdups taking place. The *News* declared that it had waited in patience until this time, but now it called upon the marshal to act.

The situation was getting out of control. Urged on by the newspapers, but needing no persuasion, A. L. White, one of the town's better-known saloonkeepers and also president of the Prospectors' and Pioneers' Association, sent a dispatch in his latter capacity to the War Department requesting the immediate assistance of troops stationed at Fort Gibbon, adjacent to Tanana. The message read: "Two hundred soldiers idle at Fort Gibbon. Gold dust to the value of $8,000,000 is without protection. Miners cannot use public highway without being held up." The War Department, with its customary delaying tactics, replied that General Constant Williams of the

Department of the Columbia had been instructed to make a thorough investigation of the situation.

Another peak of excitement was reached the following day when an organized manhunt took place. The peace of that normally quiet Sunday morning, June 25, was shattered when Tommy White, driving a stage, made his escape from an attempted holdup. Miners in the vicinity, led by Harry E. King of the Costa Roadhouse, devoted the day to beating the bush and visiting cabins on the divide and in the vicinity of the holdups. No highwaymen were found. The only result of the day's efforts were blistered feet and a higher than usual consumption at the Costa Roadhouse that evening.

One account of these events appeared in the *Seattle Daily Times* of July 16, 1905. The delay in the *Times* reporting of the occurrences was due to the disruption of wire communication. Its story was based on a Fairbanks dispatch dated June 26, carried by courier through the fire-ravaged forest country around Eagle to North Fork, 180 miles beyond Eagle, where it was placed on the wire. The Fairbanks dispatch was undoubtedly lurid enough, but the *Times* seems to have added some embellishments of its own.

The banner and lead headlines shocked Seattle. They probably would have shocked level-headed citizens of Fairbanks even more.

MILLIONS IN GOLD IMPERILLED
ON TANANA BY BANDITS

FREQUENT HOLD-UPS AND ATTACKS ON GOLD PACK TRAINS CAUSE TERROR THROUGHOUT THE CAMPS OF THE RICH PLACER DISTRICT—GENERAL FEAR THAT TOWNS MAY BE INVADED AND BANKS ROBBED.

Four days later, with communications restored, Marshal Perry replied to the *Times* story in a dispatch addressed not to the *Times* but to its rival, the *Seattle Post-Intelligencer*. He denied most of the *Time's* allegations, stated that no attempt had ever been made to hold up the treasure trains, and concluded with the astounding official pronouncement:

Al White, a prominent saloonkeeper and Wall [*News* editor] and McChesney [*News* manager] are not on good terms with the marshal's office and are inclined to be hysterical. Most of the hold-up stories originate in the *News* Office. Wires were down for some weeks and paper must be filled up, but no reason for sending yellow stories to Outside Press.

The *Evening News*, when confronted with Marshal Perry's dispatch, took exception to his entire account, calling it a distortion of events.

Meanwhile, more ominous events were transpiring in the Fairbanks camp. On June 27 the *News* reported the first shooting associated with the highway robberies. Lee St. James attempted to resist a holdup and the highwayman fired. The shot struck a silver dollar in the victim's pocket and was deflected. The highwayman fled, and St. James was assisted to Fairbanks by a chance traveler.

The *News* immediately confronted Marshal Perry with this most recent evidence of lawlessness. That same day the marshal issued a public statement that was characteristic of the indecision marking the office ever since his arrival. He stated that while doing all in his power to protect the general public, he could not use the marshal's office for police or detective work as fully as desired. There was no provision of Alaska law covering payment for such activity. He went on to say that the Alaska code provided for an arrest to be made by private persons without warrant in all respects as fully as an officer can arrest under the same circumstances. He further knew of no reason why the community could not raise money and offer a reward.

In short, Marshal Perry said that his office would not restrain people from acting for themselves, a complete repudiation of Deputy Dreibelbis's earlier warning that this kind of action would not be tolerated. The *News* labeled it a mealymouthed statement and immediately launched a subscription campaign to raise a reward for capture of the bandits. Almost overnight the reward reached $1,000.

The fury of the town represented the attitude of a vocal and frightened minority, mainly the merchants and businessmen. It was not a universal sentiment. Where one group saw a highwayman behind every spruce, another group was equally convinced that the robberies were the work of a single, daring, free-hand outlaw, the Blue Parka Man. The fears of one group were the amusement of the other. The amusement was tinged with a certain admiration that one man could take on the entire resources of the law and community, even inspire a plea for military intervention, and on occasion would even share his dangerous rewards. This gave an almost Robin Hood character to the lone bandit. Here was a man who dared, who was surely intelligent enough to know that the deck was stacked against him, but still persisted. This was high drama, excitement spiced with danger and an unpredictable climax. So the deeds and legend of the Blue Parka Man grew, a story unique in the annals of Alaska.

2

THE CAPTURE

The early days of July brought a terrifying natural disaster to Fairbanks. For a time it swept all mention of the highway robberies from the front pages of the papers and from the thoughts of the townspeople.

That spring and early summer had been the hottest and driest in the records of the new camp. The papers carried frequent comparisons of the temperatures with those of the year before. In the Alaska Range, which fed the source of the Tanana River, the hot winds were accompanied by sudden, drenching rainstorms. The usual heavy runoff from the spring thaw was intensified by the abnormal warmth and rain, with mountain glaciers melting at an unprecedented rate. The creeks were soon overflowing their banks. As they joined the already swollen Tanana, they triggered one of the most devastating floods Fairbanks was ever to know.

The level valley of the Tanana to the south of Fairbanks offered no obstruction to the swirling waters soon spilling into the town both from the south and down the Chena, which provided a natural connecting channel. The flood hit Fairbanks without warning on July 1, 1905.

By July 6, when the river crested, two-thirds of the town was engulfed by the surging water. Places thought to have been safe the day before were inundated by a foot or two of water. Every foot of dry ground was crowded with people from the flooded areas, many of whom had no more protection than a sleeping bag. A few were fortunate enough to have brought tents, which were generally crowded until the seams promised to burst.

There was no longer any effort made to transact business. Nearly every businessman facing the waterfront had a gauge stuck at the river's edge and spent most of his time watching the river climb.

After the flood receded, damage to Fairbanks' property was estimated at $100,000. As often in such cases, estimates were confined mainly to the business and commercial district. Personal losses would undoubtedly have raised the loss to a much higher figure.

At first the notice that a highway robber had been arrested passed almost unnoticed. Not until people began to forget their personal losses did they realize that the highway robberies had ceased. It was a totally unexpected development. Many people, unwilling to accept it, insisted that the search for other members of the gang be pushed. They believed this could only be a temporary lull before a new outbreak of violence. But the quiet persisted.

At last the town began to accept the fact that all the highway robberies had been the work of one man: the Blue Parka Man. With this acceptance came an insatiable curiosity. What had happened? Who was this man? There should have been, and were, many shamefaced citizens, those who had clamored the loudest for protection, who had implored the assistance of the Army. But the *Fairbanks Evening News* was not one of them. In fact, as the developments came slowly to light, the *News* was to claim major credit for apprehending the highwayman.

The arrest took place on July 11. At first only the man's name was released: Charles Hendrickson.

Immediately a score of people knew the man. Some said that it figured. Others were equally certain that Hendrickson could have had no part in the robberies, there had been some mistake. One of the most surprised men was Dan Sutherland, a miner, later to represent Alaska as a delegate to Congress for five consecutive terms.

At this time Dan Sutherland mined on No. 4 Below on Cleary Creek. Swiftwater Bill Gates, a legend of earlier Dawson days and destined to carve out another fortune on Cleary, was a neighbor on No. 5 Below. Many years

Vide Bartlett Collection in the Archives, University of Alaska, Fairbanks

Men in a placer mine

after these exciting days Dan Sutherland recalled the following incidents:

"I was personally acquainted with Hendrickson. He was a steam point man for Swiftwater Bill Gates on Cleary Creek. I worked on the adjoining claim. Our drifts were connected, and I met him there occasionally and also on the surface. He was a powerful man physically and a pleasant, jovial, and companionable fellow.

"It was said of Hendrickson that he made much more than his wages by panning rich dirt on his night shift, but that could not be substantiated. He might have done so, as point men working in the drift at night had opportunity to steal gold dust.

"The one thing that stands out in my mind after fifty years is a conversation I had with him regarding a trip to Fairbanks. It was during his operation as a highwayman that I talked with him about the trip and told him that I had a little gold dust to take in but that I was afraid of the robber who was holding up people on the trail, and I thought I would go in with the guarded pack train the following week. . . . This pack train went to Fairbanks at regular intervals, and looking back over the years I can envision John Ronan leading the cavalcade with rifle on shoulder. My surprise came when Hendrickson was apprehended, for he was about the last man in the Fairbanks district that I would have suspected."

It is unlikely that Hendrickson could have worked for Swiftwater Bill during the last month of his forays. Even for a man of his strength and endurance, he could hardly have operated so extensively throughout June and still have done a convincing job as a point man for Swiftwater Bill.

Among the others most surprised at his arrest were the volunteer librarians at Bishop Rowe's recently completed and favorite philanthropy, the fledgling Episcopal library. They recognized him as the library's most regular, courteous patron during much of the previous winter. They looked upon him almost as a teacher does her favorite pupil, the library not being exactly the focal center of community activities in the camp.

Billy Gorbracht, although not to talk about his meeting with the Blue Parka Man until years later, had probably known Hendrickson as long and as well as most. He had no illusions about the man, but no animosity.

"I never knew Hendrickson well, you might say. I don't think any man ever did. Never knew a man to call him by his first name or any nickname that was usually tagged to most men. He was a loner, and yet I never knew a man who didn't like him, and the girls, well, he was always a big favorite with them.

"It was just seven years before, in the first year of the big Dawson rush, not long except for all that has happened since, that I first met Hendrickson. We were neighbors on Sulphur Creek, where for a time I had a lay on a piece of placer ground.

"That wasn't for me. Soon I was back to music, playing at the Tivoli, and there is where I remember Hendrickson best. A blond, good-natured, big fellow. Never knew a man was ever able to lift him off his feet, and favored by all the dance hall girls. Many is the time I have seen him dance the whole night through. Lately at the Floradora I have seen a man who danced like that, but I might never have recalled him except for his stopping me that time on the trail.

"After a while I didn't see him any more around Dawson. Later I heard he had been arrested."

Al Young also remembered Hendrickson as a hard worker on 17 Below on Sulphur Creek and the details of his arrest.

"He was always the hardest worker on any crew. Never lacked for a job because he had that reputation. Whatever led him to the robbery of those sluice boxes on Hunker

Creek no one ever knew. It was a clean job, too, made off without a trace with the whole cleanup." There was an obvious trace of admiration in Al's description of the event.

"Turned out, though, Hendrickson had fooled everyone but the Mounted Police. The gold had been lugged away in a gunny sack, and that gunny sticks to a man's clothes like a burr to a horse's tail. Well, when the Mounties were taking a good look at everybody, they noticed this gunny sticking to Hendrickson's coat. That made the Mounties suspicious but didn't prove a case, because Hendrickson had buried that gold and never went near it for a year. It looked as though he had pulled it off when he returned a year later to dig up the gold. They let him do the whole job of digging, too, and when the sack was uncovered and Hendrickson looked up, there were the Mounties covering him. They must have watched his every move for that whole year. He got a five-year sentence, and that was the last I ever heard of him until he was arrested as the Blue Parka Man.

"Was I surprised? Not really, I guess. He was always a bold one. It fitted his style.

"Did I tell you he never offered or accepted a drink? A fact. Never touched a drop." It was as though this final statement was the ultimate explanation for Al.

Tom Gibson, who had good reason to remember Hendrickson later, describes him most graphically: "He looked like a round-faced Swede, light complexioned, with light blue eyes. The fellow spoke with a slight accent—Swede, Norwegian, anyway some kind of Skewegian. He was I guess about thirty-five or so, his hair very bushy. He was not tall like the slim Swede but above middle height and well set up like perhaps a fighter with those rippling muscles. I never knew one quite like him. He was always jovial, talked a streak as though he never had a care, and yet you felt you could never reach him."

Finally, a sketchy account of Hendrickson's arrest was released. Everyone knew the marshal was not telling all the story, was holding back the most incriminating facts for the trial. Most of the details of the arrest came from the few eyewitnesses. Bill Butler, one of these, recalled the events many years later.

It was midmorning of a Tuesday, July 11, 1905, when Charles Hendrickson entered the Northern Commercial Company to order some supplies. What aroused the clerk's suspicion will never be known. The Blue Parka Man's physical description was well advertised. That could have been enough, sparked by a desire for the reward money (which the clerk was never to receive). At best it could have been but a slender hunch. On the pretext of having to get some of the supplies from the warehouse, the clerk left Hendrickson and called the U.S. marshal, whose deputies were already on the prowl for Hendrickson.

Hendrickson waited for a time. When the clerk failed to return, he became suspicious and left the store with his order unfilled. He had covered only the short distance to Third Avenue opposite the Fire Hall when he was met by Chief Deputy E. R. Reynoldson and Deputy Marshal George Dreibelbis. The arrest was totally lacking in the dramatics usually associated with the showdown between frontier marshals and desperadoes. The marshals had the drop on Hendrickson, who made no effort to resist and was promptly lodged in jail.

Although lacking in theatrics, it was a fateful encounter, this first meeting between Hendrickson and Dreibelbis. They were to become the chief actors in one of the longest, most deadly manhunts in the annals of the North. Neither man on this first meeting could more than have glimpsed the mettle of the other—the bold cunning

and imagination of Hendrickson, nor the steady, inde-
fatigable persistence of Dreibelbis.

Dreibelbis saw a man whose slightest movement revealed rippling muscles straining against his gray woolen
shirt and tightly fitting Levis, while the near-noon sun
burnished his yellow hair the color of Cleary Creek gold.
Hendrickson was watching and smiling at Dreibelbis,
who stood a good four inches taller, thin and with trail-
hardened muscles that hardly showed, a complete contrast in physical makeup.

The law was determined to mete out quick justice to
Hendrickson. Already the short mining season was half
over, and Hendrickson had disrupted a good part of it. He
had humiliated the marshals and raised a fury that carried
all the way to Washington, D.C. The wheels of the law
lost traction in an unexampled burst of speed.

Almost immediately after his arrest, three charges were
lodged against Hendrickson: larceny, obtaining money by
false token, and highway robbery. Only nine days from
the date of his arrest, he was brought to trial on the first
count of larceny. The prosecuting attorney presented an
imposing list of four witnesses in the supposedly cut-
and-dried case. Hendrickson, represented by the law firm
of Heilig and Tozier, called but a single defense witness.

3

THE TRIAL

The morning of the trial dawned clear, hot, and brilliant, after the sun rolled briefly under the rim of encircling hills and started again its twenty-four-hour arc across the northern sky.

Much of the town had spent the night in the saloons to be sure of an early morning start. Hours before court convened, people filled the street in front of the frame courthouse. Only a few found seats or standing room when the courtroom doors finally opened, but those crowding the open doorway passed on to the eager throng still lining the street behind them each detail of the swiftly developing trial.

The prosecution lost no time in presenting its case, and it was one of the most bizarre stories handed to a Fairbanks jury. First, the man who initially identified Hendrickson to the marshals was put on the stand. It proved to be Officer Hayes of the city force, attracted by the reward offer, who had hired a substitute in his place on the force and took up the trail that led to Hendrickson. There were angry murmurs in the courtroom, and the words "stool pigeon" were repeated by several. Already the mood of the spectators was evident, and it must have made the prosecution apprehensive. If this start was disconcerting, what followed was even more so. The *Seattle Post-Intelligencer* in its July 20, 1905, issue carried the following story:

C. Hendrickson was arrested a few days ago on a charge of burglary. The accusation was today changed to highway robbery, and he will be tried on that charge. Back of this alteration is a queer story of life in the far north and some reputed queer notions of a "square deal" and "tainted money" about which so much is heard in the States.

According to the information brought out at the hearing here, Hendrickson was given away by some of his countrymen on account of a breach of contract, which recalls the days of Robin Hood when the rich were made to contribute involuntarily to the support of the poor. Hendrickson, it is charged in the information filed, entered into an agreement by which he was to have the moral support of some of his "friends" in robbing the rich operators, but he was not to molest the poorer men. Hendrickson is accused of failure to observe this limitation and to have poked his gun under the noses of some of the mining laborers. This was too much for human nature, and a warrant was sworn out for his arrest. . . .

The prosecution could hardly have presented four more dubious witnesses, undoubtedly rounded up by Officer Hayes to support this remarkable conspiracy. As one uncomfortable witness followed the other to the stand, the mood of the spectators changed from sullen resentment to open hilarity. This was a better show than one paid to see at Century Hall.

It was obvious now even to the prosecution that it had committed a serious error: not only had it lost the present case, but it had prejudiced all future cases against Hendrickson. The prosecution hurried through its summation. The sooner this was over, the better. By noon the case was in the hands of the jury.

The jury, not to be outdone by the prosecution, reached its decision by midafternoon. At the court's direction, the jury foreman arose and delivered the verdict everyone was expecting.

"The jury finds the prisoner not guilty."

As though a sluice gate was opened, pent-up sound swept the courtroom, drowning the frantic banging of the judge's gavel. Wild cheers merged with hoots of derision. Everyone was on their feet at once except the prosecuting attorney, still too shattered to get his feet under him.

George Dreibelbis, a silent spectator of the entire drama, saw the impossibility of quelling this outburst immediately. He made his way through the crowd until he reached the prosecuting attorney. His hand on the attorney's shoulder was like waking a man from shock. The attorney jumped, then lunged toward the judge's bench. There was a shouted conversation, indistinguishable in the din, before the judge began pounding his gavel once more. At last the crowd was willing to listen. As the court stilled, the judge's voice, hoarse with anger, pronounced: "This prisoner will be remanded to jail to await trial on the other counts against him. Now clear the court."

Many of the spectators had figured that a non-guilty verdict meant freedom for Hendrickson. But even with the frontier juries Dreibelbis knew so well, the law still held the advantage.

In the days that followed, the prosecution settled down to a serious job. There were frequent consultations between the court and the district attorney. In fairness to the forces of the law, they had acted on sincere conviction. To them the circumstantial evidence was overwhelming,

and it continued to mount. Marshal Perry soon received a
reply to his earlier query to the Northwest Mounted
Police, dated Dawson, July 26, '05:

Dear Sir
Inclosed please find the photo of Carl Hendrickson whom I
have been informed you have under arrest for hold up.

He was arrested here and convicted June 13th '98 and sentenced to 5 years H.L. He is a bad thief and a bad man and I
am in hope you will be successful in handling him and by
doing so you will do away with a bad man. When you are
through with the photo please return it. I know about 50 ex
convicts that are in your country that have done time here or
out side, and any time I can be of any assistance to you let me
know.

> Yours Respt
> W. H. Welsh
> Dex

The photograph positively identified Charles Hendrickson, but as with most of the evidence against him, it
did not identify him as the Blue Parka Man. This was the
frustrating part of the case for the prosecution. The most
damning evidence against Hendrickson was that all robberies had ceased since his arrest. But this was not admis-
sible evidence any more than it was proof. No witnesses
had come forward who could positively link Hendrickson
with the activities of the Blue Parka Man. This feeling of
helplessness in the face of massive circumstantial evidence had led to prosecution on the word of some less
than exemplary citizens. Frontier juries were not obtuse;
they merely wanted a man to be given a square deal.

The lesson of the verdict was not lost on either court or
district attorney. The grand jury was in session, and the
district attorney chose not to prosecute on the second
charge of obtaining money under false pretenses—

Laura M. Hills Collection in the Archives, University of
Alaska, Fairbanks

Main Street of Fairbanks, 1904

allegedly cashing a check with no money in the bank.
Instead, win or lose, the district attorney went to the
grand jury with the main issue of highway robbery, and
on this charge obtained an indictment. The new trial was
set for early August.

Undoubtedly the temper of the town also had much to
do with the district attorney's decision. Once again the
newspapers were having a field day at the law's expense.
The *Evening News* now claimed all credit for Hendrickson's apprehension. Without their offer of the reward and Officer Hayes's sleuth work, they claimed,
Hendrickson would still be on the loose. The marshal's
office came in for another round of abuse for its alleged
inaction during the robberies. Not content with this
achievement, the *News* now took the lead in ridiculing

the conduct of the trial. At least the marshal must have found some comfort in the fact that the main attack was now directed at the district attorney and that he, mercifully, was a secondary target.

The *News* seems only to have mirrored the temper of the town, or one segment of it. There were many, like the *News*, who were infuriated at the ineptitude of the district attorney and the failure to secure a conviction. But there were others who felt just as strongly that Hendrickson was the victim of a contrived conspiracy. Everyone was taking sides in the mounting drama. If anything, Hendrickson had gained strength. Many people, neutral before, condemned a second trial as legal technicality in face of the acquittal. One thing seemed certain: Unless a much stronger case could be presented, Hendrickson stood to come off scot-free.

There is nothing but hearsay to connect Hendrickson with Gabriel. She may or may not have been his girl. Perhaps the incident on the row the night of Hendrickson's trial stemmed only from the violent partisanship that the trial evoked, or perhaps it was actually a case, as some contended, of revenge for betrayal. The *Fairbanks Evening News* recorded the event.

Judge Erwin yesterday imposed a fine of $50 against Gabriel, the French woman who started a fight on the row with Irene Wallace and pummeled and beat her face so bad the services of a doctor were required. The French woman graphically described to the court how she delivered the blows which did such terrible execution upon her opponent's face. She admitted that she knew something about the art of prizefighting and punched holes in the air to show how different blows could be delivered with force. The evidence showed that she had tried to get at her opponent an hour after the fight to finish her. The Wallace woman evidently was in great fear of her life, for shortly after the trial she was aboard the boat bound for Nome, and left Gabriel the victor, as well as the terror of the row.

The grand jury had other business before it in addition to Hendrickson's indictment. As a part of its normal functions, it investigated the conditions of the jail and delivered a severe rebuke. Its report was carried in the *Fairbanks Times* of August 3.

On the rear of the federal building lot is a small log building used for a jail, which is more of a dungeon than a house of detention, being able to accommodate fourteen persons in three rooms only, and yet at times housing twice that number. It is wholly insufficient for the size of the Fairbanks district at present. Some immediate relief from these dank, close quarters may be had by putting in a skylight and ventilated closets; and by giving the prisoners daily outdoor exercise.

Judge Wickersham in his diary described the jail as a poorly constructed log house built in 1903 at a cost of $1,750 when labor and material were both scarce and high. It was thirty feet long and twenty-four feet wide, containing a guardroom, corridor, and three cells.

This was the place in which Hendrickson was confined. His second trial would not be long coming, and that was good. After this case was settled perhaps the town could cool down again.

George Dreibelbis had taken to visiting Hendrickson daily. This was no part of his job, any more than the close scrutiny he kept on all of Hendrickson's visitors. There were not many. Leroy Tozier, Hendrickson's lawyer, was the only consistent visitor, and he always exuded an air of overbearing confidence. Dreibelbis could not put down his feeling of uneasy concern and, he admitted, a growing curiosity about Hendrickson.

On the occasion of this particular morning's visit, Dreibelbis found Hendrickson absorbed in the ritual of a two-handed pinochle game with the cell's only other prisoner at the time, Paul Burkall, a petty criminal charged with dog stealing.

Hendrickson pushed his chair away from the bunk, arose, and walked to the cell door. He greeted Dreibelbis with a grin.

"How about doing me a favor? Get me some books from the library. The lady there knows what I like. If the kids are through with the old funny papers, bring them along for Paul here."

Dreibelbis nodded and turned to go.

"Say, get me some tobacco, too. My credit is good." Dreibelbis didn't pause. "Here, catch this," added Hendrickson, flipping a five-dollar gold piece through the bars at Dreibelbis's feet. "Just in case your credit isn't as good as mine," he said.

Dreibelbis retrieved the coin. Now how did the jailers in their search ever miss that?

Dreibelbis's daily visits to the jail continued, and each day he brought two new books in exchange for the ones Hendrickson had finished the day before. Heavy reading much of it, lots of stuff on geology. It puzzled Dreibelbis for a time, until he remembered hearing that Hendrickson was a graduate engineer. In fact, as the days passed, everything about the man seemed irreconcilable. Hendrickson was a maze of contradictions—a highly educated man who chose crime for a livelihood, a hard worker who preferred to play the odds, a popular man if he had wanted to be but solitary, a confirmed criminal but not hardened, a man who robbed but could show generosity, an intelligent man but totally indifferent to his own interests. And strangest of all, he was a man who had succeeded in effacing his identity.

The trial was only a matter of hours away. Dreibelbis felt more relaxed when he entered the jail on the early morning of August 8. He was the first man to discover that Hendrickson had escaped, along with Paul Burkall.

Dreibelbis roused the jailers, who hastily unlocked the cell. The scant bedding had been arranged to resemble two men still asleep in their bunks, but the gaping hole in the fourteen-inch–thick log foundation beneath Hendrickson's bunk told another story.

Dreibelbis wasted no time in further investigation. He had learned enough to know the prisoners could only have a start of two or three hours, and he was soon heading a posse in pursuit of the fugitives.

Hendrickson, using a small file, had fashioned a saw out of an iron hoop. It was an ingenious piece of work that would have done credit to a machinist. It had to take long hours to make, to say nothing of the time required to saw through the fourteen-inch log. He could have had little time for sleep during the days of his confinement, although undoubtedly Burkall had been put to good use with the actual sawing. How Hendrickson had obtained the file and the iron hoop became a favorite topic of speculation. Some said they could have been passed to him through the moss-chinked cracks between the logs of the jail. The opinions were almost endless.

Hendrickson and Burkall were seen at nine o'clock on the morning following the escape. Frank Sickenger and J. Miller encountered them just beyond the Owl Roadhouse. Burkall was in the lead, perhaps a hundred yards ahead of Hendrickson. He stopped Frank Sickenger, whom he knew slightly, to ask for a dollar to eat on, then moved along. When Sickenger and Miller met Hendrickson moments later, they had no suspicion of his identity. That only came later after the two men reached town. Hendrickson was carrying a brand new .30-.30 rifle; he stopped and rested the butt on the ground, with his hands over the muzzle. He matched the general description the men had heard of Hendrickson, but he had shaved off his mustache. They talked about the condition of the trial as casual strangers will, but Hendrickson kept glancing back along the road and soon continued on his way. At the Owl Roadhouse, Sickenger learned that Hendrickson had stopped there briefly, leaving his rifle outside resting against the cabin.

The meeting with Sickenger and Miller was the only sighting of the escaped prisoners in the days to follow. The posse returned to town empty-handed. The newspapers were incited to new heights of invective. The marshal's office reached a new low in public opinion. The district attorney must have been slightly relieved, for at least he had a case he could now prosecute with assurance of conviction—if Hendrickson was ever apprehended again. If anyone could have felt as low as the forces of the law, it was probably Leroy Tozier, now that both his client and his fee had escaped.

Probably Judge Wickersham best expressed the frustrated feeling of the law when he wrote in his diary on August 13:

> Tried Fred Owens for murder of Carl Christiansen at Forty Mile in November 1901, all week. Verdict yesterday; "Not Guilty." This is the second radically bad verdict at this term—Hendrickson and Owens were both viciously guilty, but soft-hearted and soft-headed jurors acquitted them. Hendrickson sawed a log out from underneath his bunk—a day or so ago—and escaped—but it was not necessary. He had better trusted the jury!

INTERLUDE

After Hendrickson's encounter with Sickenger and Miller, there was only one other uncertain report of an encounter. This vague sighting occurred far up the Chena Valley. After that there was nothing. It was as though the wilderness had swallowed the man. Once again heavily armed guards were placed on the gold trains. Solitary miners, prospectors, and travelers wasted precious days until a group could be assembled headed in the same direction. Only the foolhardy ventured the trails alone. Much of the amusement of the early summer vanished. The quarry was known now, and few doubted that Hendrickson would be willing to trade his life with any man at any time to avoid recapture. Although the robberies were not resumed, there was no lessening of the vigilance or of the fear that gripped every man on the trail.

To Dreibelbis the possibility that the wilderness might indeed be the end for Hendrickson was simply too easy a solution to accept. Not that it was an idle possibility. The wilderness had trapped many a man before and would claim many more.

All the odds were against Hendrickson's survival. The

Chena lowlands were a living hell that late summer and fall. The slowly receding waters from the spring flood had left every marsh and slightest depression a stagnant breeding ground for swarms of mosquitoes, the monotony of their unceasing buzzing enough to drive a man mad. The pests found every unprotected spot—ankles, wrists, an empty buttonhole, a snag in a pair of Levis. Even the indispensable head net became an aggravation rather than protection when a man's sweat, as he trudged through muskeg and over deadfalls, plastered the net flat to his face and allowed the mosquitoes fresh opportunity for attack. Worse followed when the first trace of frost brought relief from the mosquitoes, for then the angry hordes of gnats appeared, vicious insects that bit rather than stung, raising purple welts that burned with a wild, crazy fury. It was often a losing battle that man fought under these conditions—his judgment fraying with the stabbing insect bites and his legs seemingly turned to putty with each plodding step through the quivering, spongelike tundra.

All this Dreibelbis knew only too well, and he knew also that Hendrickson had to be traveling light and now probably alone. He would not allow himself to be impeded for long by Burkall; his was a solitary pattern.

What would be Hendrickson's next move? It was an unanswerable question. After he eluded the dragnet thrown up following his escape, there had been no further attempt at pursuit, and the posse was disbanded. If Dreibelbis had learned anything from his experience in the bush, it was the folly of pursuit under such conditions. The hunter soon became the hunted in that kind of game. The wilderness was the lawman's best ally. If Hendrickson survived, and Dreibelbis surmised that he would, starvation and the onslaught of winter would inevitably drive him back to society.

For Dreibelbis, however, waiting was the toughest course he could have set for himself. Even worse than the constant public criticism were Marshal Perry's occasional inquiries about his progress. Early on Marshal Perry had recognized the strange intensity of Dreibelbis's feelings. Hendrickson's apprehension had taken on an unexpected dimension that was to lock the two men in a duel to which the law was only an incidental spectator. If Perry had any qualms at this development, they were quickly buried in a feeling of relief. Hendrickson became Dreibelbis's personal mission.

Dreibelbis admitted a fascination about his man that he could never satisfactorily explain. Perhaps in Hendrickson he recognized a free spirit that made him akin to the wilderness and so an opponent to be respected. But it had to be more. The unvoiced challenge in Hendrickson's attitude at the trial had rankled Dreibelbis. Even beyond all this, however, were the many paradoxes in the man's nature that Dreibelbis could not understand. And finally, when all the other questions had been asked, Dreibelbis always found himself reverting for the thousandth time to the most unfathomable of them all: Why had Hendrickson chosen the wilderness to the jury?

In the days and weeks that followed, Dreibelbis methodically developed his plan. He met and questioned every man and woman who entered town. Usually these were casually arranged meetings so they appeared to be more an exchange of gossip than any formal questioning. From these apparently chance encounters Dreibelbis developed an encyclopedic knowledge of every near and remote settlement, their occupants and businesses, the miners and their crews, the solitary prospectors and the trappers. He learned of all the comings and goings, since every stranger was news in these activity-starved oases of men. He tried not to grow impatient at his plodding progress, for he knew that with each fresh shred of information he was baiting a trap into which his man must inevitably wander. In the process he could not overlook the people of the town.

Dreibelbis was haunted by memories of many unexplained occurrences. There was the flash of the golden half eagle spinning at his feet on the jail floor where Hendrickson had tossed it. After that Dreibelbis had made certain that Hendrickson was thoroughly searched again. Nothing was found. Yet Hendrickson had acquired the file and the iron hoop with which he effected his escape. When seen at the Owl Roadhouse he was equipped with the new Savage rifle and with trail gear. Did Hendrickson have friends in the camp who assisted his escape? It was a strong probability that demanded investigation. Dreibelbis had set a tedious job for himself, and he felt the strain as the last days of August—another rainy, miserable August—passed with no tangible results. The trail seemed cold as the grave.

When the gloomy August rains finally did end, almost immediately there was a touch of early winter in the air. The sun now shown brilliantly during the rapidly shortening days, and in the dawning hours, the first traces of frost sparkled like iridescent jewels from every leaf and blade of grass. On the trail a man awoke to find the moisture of his night's breathing frozen in a glittering necklace on the rim of his sleeping bag, and his steps for the first hour were like walking on cement. Then the sun, hot and brilliant, dissolved the frost and briefly recalled summer until the evening shadows settled.

The riot of fall colors also seemed to appear overnight: scarlet buck brush, crimson cranberry and wild currant bushes, and purple wildfire mingled with the dazzling golden glow of the birches amid stands of tamarack and spruce until the eye was blinded by the explosion of color. It was as though Nature in these short, unbelievable autumn weeks sought to rob man of his senses, to promise a fulfillment that would lull him into imagined security before the onslaught of another arctic winter. At last even the gnats were gone, and the game was sleek and fat. The land was suddenly and briefly a veritable paradise.

On September 23 the Fairbanks headlines blazoned the latest news of the Blue Parka Man. Hendrickson's cache, from which he had presumably operated at the height of his forays, had been discovered.

Bob Roberts, a Fairbanks resident, found the site purely by accident while hunting near the Gilmore Roadhouse. He described it to the reporter for the *Fairbanks Weekly News* as an impregnable stronghold. "The place is almost inaccessible. The lay of the country is rough, and big granite boulders are thrown up in all kinds of shapes. One man could pick off fifty men from this retreat, or he could play a game of hide and seek among them for a long time."

The location was near the head of Nugget Creek, about two and a half miles from Gilmore. There, between two big boulders, Roberts found a blue drill parka, a blue fly tent, partial sacks of flour and rolled oats, a frying pan, a blanket, and a pair of old overalls partly mildewed from the damp.

Immediately speculation was rife concerning the present whereabouts of the Blue Parka Man. He was rumored to be again in the vicinity, and it was suggested that only the heavy August rains had prevented resumption of his attacks. The community was warned to expect robberies as soon as the ground froze hard enough to permit easy travel.

The same issue of the *Fairbanks Weekly News* that broke the story of the discovery of Hendrickson's cache carried another interesting sidelight on the events of the past summer. Officer Hayes of the city police force, who had taken leave of his duties to carry on the undercover search for the Blue Parka Man, had become justifiably impatient at not being paid the $1,000 reward offered for the highwayman's capture and conviction. Some subscribers to the reward had even reneged on payment, for a variety of reasons: no conviction, Hendrickson still had to be proved guilty, and probably most compelling of all, there was no prisoner! It was finally decided, and duly reported by the *Weekly News* September 23, that the Washington-Alaska Bank should turn over to Officer Hayes 10 percent of the subscription collected to date. The sum amounted to a munificent $83.50. There is no doubt that many citizens considered it just payment for clandestine snooping into another man's affairs.

5

THE CARIBOU HUNTERS

The current court session came to a close in late September. There was a sharp tingle to the early morning air now. The first flurries of snow melted almost as they fell, but there was little doubt that the next snowfall would remain until the following April. Throughout the North that year, winter settled in unexpectedly early. On September 22 ice froze at Whitehorse in the Yukon to a thickness of half an inch.

What had happened to the summer? The fantastic blooming of the arctic summer was always brief, but this year it seemed to have been only a promise. It had been just eleven weeks since the July floodwaters started to recede; August with its unrelenting rain did not deserve to be counted, and already the hawk was screeching.

It could not be better for his own purposes, Tom Gibson reckoned. He had received word that the caribou were running at last, and now the kill would freeze quickly and keep indefinitely. It was rarely so easy. Often the caribou runs began while the last traces of summer still lingered. Then the caribou had to be quickly skinned and cleaned, hung in as airy a spot as possible, and liberally sprinkled with cayenne pepper to protect the meat from the flies until a firm crust formed. Then one prayed there would be no more rain to sour the meat. Now, however, with the blessed freezing temperature, the carcasses, after cleaning, could be stacked like cordwood until they were ready to be hauled to market. Once in a while a fellow lucked out.

The word reaching Tom had come from the brothers Frank and Chuck Larson, his hunting partners that winter of 1905–1906. Together they had left Fairbanks in early September for Tom's cabin at the head of McManus Creek, just across the divide from the Twelvemile Roadhouse at the head of the Chatanika River. Two short years before, in the first flood of the Fairbanks stampede, this had been the main route of the stampeders in their frenzied rush to the new Eldorado. Roadhouses sprang up at twenty-mile intervals to accommodate the travelers. The land was suddenly peopled by a swarm of invaders

Clara Rust Collection, Archives, University of Alaska, Fairbanks

Home of a trapper

Clara Rust Collection, Archives, University of Alaska, Fairbanks

Mining area near Fairbanks

but, swiftly as they came, they departed. With the opening of the much speedier, more convenient water route to Fairbanks, the roadhouses were abandoned. Only a few stubborn or tired owners hung on in the hope of a new miracle.

For the most part, the land reverted again to its original inhabitants before the coming of the first man: the upland caribou. Each fall the animals banded together in herds numbering in the tens of thousands for their migration from the White Mountains calving grounds to their southern winter range. Tom Gibson had selected his cabin site well, in the center of this age-old migration route.

A busy month for Tom and his partners had preceded Tom's brief return to Fairbanks. Much of the winter's supply of dry wood had been cut and stacked for the gluttonous Yukon stove that provided the cabin's only heating and cooking facilities. The cabin itself had needed some rechinking, and a new room was added to provide more adequate space. With most of these winter preparations complete, Tom left his partners to finish the wood cutting while he returned to town. Arrangements were quickly made with Pete Malone, the freighter, who on the first good snow would deliver the remainder of the partners' winter supplies with his horse-drawn freight sled and then begin freighting the fall and winter kill back to the meat-hungry town.

The excitement of the hunt began to flow anew through Tom's veins as he reread Frank's note telling him of the caribou migration. He had just twenty minutes to catch the Cleary stage. Through his second-story hotel window he saw snowflakes lazily starting to fall, each flake seemingly as large as a two-bit piece. Soon they would dwindle in size, then start falling so thickly that it would appear a curtain had been quietly drawn across the landscape. Yes, it was time for the snow to stick now.

Briefly, Tom recalled his meeting of the night before with George Dreibelbis. Between old acquaintances there had been no need to beat around the bush. No, Tom had heard or seen nothing of Hendrickson. That was to come later.

At Cleary City, Tom picked up the two horses the partners would need to yard up their caribou kills. Left to freeze for a couple of days, the carcasses could then be easily moved to a convenient location near the cabin, there to await the arrival of Pete Malone and his freight sleds. Saddled and trailing one horse, Tom was soon on his way, but the going was slow. The icy, rutted trail demanded caution.

Anxious as Tom was to arrive at the hunting grounds, he had one ritual he must perform first, and it had to be taken care of immediately now that the caribou were moving. That was his visit to Fred Nichols's Twelvemile Roadhouse. Nichols was one of the relics of the earlier stampede, better long ago buried but too cantankerous to oblige. Nichols's roadhouse was one of the few still in existence on the old Circle City–Fairbanks trail. Located just over the Twelvemile Summit from Tom's cabin on McManus, the two were close neighbors—but not from choice. Their feud was long-standing. They had quarreled many years before in Dawson, more recently in Fairbanks. But the crowning insult for Tom was that Nichols expected a cut-price for fresh-dressed caribou. Right then, Tom promised himself that Nichols would never again have a chance to enjoy the savory caribou steaks he was too stubborn to buy at an honest price. That was not as difficult a promise for Tom to keep as it might seem, even with the country overrun by caribou. Nichols was one of those front-porch Alaskans like many others—take them a step into the brush and they were lost—and it was easy to keep the caribou out of his range, limited as it was to his doorway.

With such pleasant thoughts Tom passed the time as the long miles and hours fell slowly behind. Frequently he had to dismount and walk to restore circulation, stamping numbing feet and flailing his arms back and forth. He had to spend two nights on the trail, lucky enough on one occasion to find an abandoned cabin with the luxury of a stove, but on the other having to make do with only a cabin roof. Dusk of the third day found him near the Twelvemile Roadhouse, and with no trace of caribou in the vicinity. Gratefully he swung the horse's head over the ridge on the downgrade to warmth, food, and a comfortable night's sleep.

Hunter with caribou

The following morning the partners decided to spread out and hunt alone. The growing Fairbanks market was crying for meat, and the prices were the highest Tom had ever known. The thought of those prices kept Tom going in the face of the changing weather. Toward midafternoon gusty winds and swirling clouds began creeping across the high divide, turning the land into a forlorn, demented world. Tom had about decided to call it a day and head back to the cabin when he spotted a large band of caribou. Quickly he jacked a shell into the chamber of his rifle. His first shot dropped the leader and set the herd to milling. Making every shot count, he killed eight of the fattest animals before the band was out of range.

His reaction had been almost automatic, and he was soon cussing himself that he had not let the animals go. The cleaning and dressing in the fading light took longer than he had expected, and his hands were burning from the cold. Momentarily, as he opened the belly of each animal, the hot blood restored a measure of feeling to his fingers, but almost immediately they were colder than before. As his hands became progressively clumsier, it took longer to clean each animal. When the job was finally done, he discovered that fog had completely blotted out the landscape. He groped stumblingly each foot of the way to the summit. Luckily his partners would have reached the cabin well ahead of him, would have the coffeepot going and thick caribou steaks sizzling on the stove.

Tom did not notice the figure looming out of a swirl of mist until a shoulder almost brushed him. The man carried no pack, but Tom, trained to notice such things, saw the .30-.30 Savage on cock, gripped in the left hand. Despite the shadowed light and the mist swirling between them, Tom was uncomfortably certain that this was the Blue Parka Man. He had been one of the lucky few to find

standing room at Hendrickson's trial, and now he wondered if his luck might have abruptly ended.

He topped the stranger by a good four inches, but that
powerfully muscled frame offset any height advantage.
Tom did not carry a shell in the chamber of his rifle, so he
was outweighed there.

"Where you from?" Tom asked. "You must be lost, else
you're another market hunter or crazy. No difference
there."

The man grinned easily. "Twelvemile Roadhouse," he
answered.

"Come on. It's only a short way. I'll put you on the trail.
From there it's all down hill. You can't miss it."

"I'm taking no chances in this weather," the stranger
said. "You must have a cabin nearby. How about spending the night with you?"

"There are three of us bunked in that cabin. Sardines
are more comfortable. There's no bedroom."

"Hell, the floor is good enough for me. Lead the way."

There was no use arguing, Tom decided, but he would
be damned if he would walk with a cocked rifle at his
back. The Savage, he noticed, was the kind with a little
square on top. This square was in an up position when
the gun was ready to fire, and it was ready now.

"You are not stumbling along behind me in this fog
with your gun on cock," Tom growled. "Put it on safety
and we go."

The man's face broke into a broad smile. He flicked a
glance at his rifle.

"Well, it sure enough is." He flipped the safety. "Can't
be too careful with these things, can you? That's one of
the new Winchesters you're carrying, isn't it?" His gaze
was admiring. "First one I've seen. How do you work it?"

Tom worked the mechanism fast, so fast he hoped the
fellow could not see how it operated. This, he sensed,
could be a damned unhealthy interest.

"Just follow," Tom said shortly.

How did he get there? Tom wondered, acutely aware of each footfall behind him.

It appeared certain now that Hendrickson must indeed have struck up the Chena River Valley following his escape, then up the Little Chena and probably across the divide to Smith Creek. From there his trail would inevitably have taken him to its junction with the Chatanika. Directly north of the Chatanika at this point lay the high plateau country drained by McManus and its sister creeks of Faith, Hope, and Charity, where Tom and Hendrickson now confronted one another.

It was all logical enough to a point, that point being the stranger's ghostlike emergence from the fog at Tom's elbow. That and his appearance, for he was wearing what appeared to be secondhand but serviceable trail clothes: wool pants, heavy mackinaw, wool socks, and trail moccasins—not the best fit and certainly recently acquired. Moreover, there was absolutely no accounting for his presence at the time and place of their encounter.

Conscious of the footfalls on the crusted snow behind him and the uncertainties of his situation, Tom tried to fit together the pieces of this strange puzzle. When the stranger had said that he was staying at the Twelvemile Roadhouse, it was no doubt the truth, for no other answer would have been as probable. But that still left almost as many questions unanswered as before.

When he judged the cabin to be close, Tom let out a bellow at the top of his lungs. Somehow he had to tip off his partners, but he was not sure how to manage that. At his third yell a blaze of light loomed up ahead, and there, thankfully, was Frank framed in the low doorway. If it had not been for the light pouring out of the wide open door and Frank's cheerful shout, they could have passed ten feet from the cabin and never seen it.

"There it is," Tom said to the man at his shoulder.

"Shove on in. Frank, we have company, but get out here and help me off with this pack."

Quickly Tom grabbed Frank by a shoulder and pulled him outside, at the same instant reaching behind to half shove the stranger through the doorway. As he ducked to enter, the light fell squarely on the man's face. Frank, who had started to object the instant before, stopped in mid-sentence. In the second before the door swung shut, leaving them enveloped in total darkness, Tom had a glimpse of Frank, his face a picture of stupefied amazement.

"Christ, Tom, that's the Blue Parka Man," Frank gasped.

"I know. Shut up and listen."

Frank was the oldest of the two brothers, but Tom knew that surely did not make him the smartest.

"There's a big reward for him," Frank whispered, "and there are three of us."

"You damned fool. He is not our business. He leaves us alone, we do the same. I got no hankering to bury you or him. Now get inside and try to act natural."

Chuck was already slapping the juicy, steaming, thick steaks on the plates when Tom entered. A man could taste the rich, savory flavor with his nose.

"Wash and pull up," Chuck ordered happily. "We weren't holding dinner for any fools out strolling on a night like this."

It was the kind of supper Chuck could fix better than anyone Tom knew. He dished up thick golden gravy, steaming rice, steaks, bannock, and good boiled coffee to wash it down.

Chuck talked noisily and more than usual throughout the meal, so Tom figured he knew the score without being told. Frank was quiet, knowing he could expect no help and not anxious for any solo effort.

Hendrickson did not say much until he had wolfed down his third helping. He ate with the single-minded concentration of the wolf in its first kill after a long hunger. Tom could remember times when he had been like that himself after long weeks of eating off a dwindling grub pile that, without the delicacies of coffee, salt, and sugar, was as tasteless as caribou moss.

After wiping up the last trace of gravy on a piece of bannock, Hendrickson made it an evening never to be forgotten. He sang and related stories that convulsed the partners, recalling events and people of the Dawson days with a born storyteller's color. It was as good as a ten-dollar vaudeville performance, even better because it was all of familiar scenes and faces.

Tom, watching and enjoying it all, thought that if it was not for the rifle always so close to his hand, one would swear that the man never had a worry in the world. He was a sure enough puzzle. The eyes, together with the ever-handy rifle, were the only contradictions to the jovial, entertaining guest, for the man's laughter never reached his eyes. They were not mean or flat like a killer's eyes, but wary and solitary, with an indifference that sealed out all emotion that touched the average person.

Tom was not one to be easily embarrassed if found sizing up a new acquaintance with more than common interest. This was a usual habit in the North, where a man's fiber was more important than false sensibilities. Not that it would have mattered in this case in any event, because he knew that all three partners were being sized up with equal intensity by their guest. And now Tom was convinced beyond the least doubt that the stranger was indeed Hendrickson and, moreover, that Hendrickson in turn knew he had been recognized. Beneath the hilarity of the evening was wariness, assessment, speculation, and finally a guarded, unspoken truce. It was as though they were all captives of the wilderness for the moment.

Tom was certain that he came to know more about Hendrickson that night than he was ever to know about any man in so short a time. From the way he wolfed his food, Tom knew he was only recently on the verge of starvation, a man for whom food still possessed a life-and-death quality most men never experience. How long since he had ended his self-imposed exile? Ten days, two weeks perhaps? That was a bit difficult to say. One ear was just returning to normal after an obviously severe case of freezing. From time to time Hendrickson would rub the lobe of the ear, and shreds of dead skin would peel off at his touch. A still livid scar extended across the bridge of his nose and one eyelid, where he had undoubtedly been whipped by a rebounding willow branch that must have left him half blind for several days. These were unmistakable badges of the wilderness. And there were others: his gaunt though powerful frame, still hollowed cheeks, stomach flat as a board, and a rawhide toughness revealed by every movement.

All this Tom noted, plus other even more important qualities of the man, if things were ever to come to a showdown. Hendrickson was slightly hard of hearing, a fact he tried to hide, not quite successfully. He was somewhat nearsighted. It was as though Nature had endowed the man with such a tremendously powerful physique that it had to compensate for its largess with these minor, almost indistinguishable flaws.

Still unanswered were the questions of where Hendrickson had obtained the winter garb he was wearing and his presence in the fog atop Twelvemile Summit. The light clothes he wore on his two-month-long trek up the Chena would have been threadbare by the time he reached Fred Nichols's Roadhouse. There he could have obtained the winter clothing, and from the way Hendrickson's body strained at every seam, ready to pop the shirt buttons, Tom judged that the clothes were some of Fred's spares.

Hendrickson must have paid dearly for whatever he received, because one thing Nichols was not was a philanthropist. For another, Nichols was the nosiest critter Tom knew, and Hendrickson's appearance on arrival would have aroused his endless speculation and questioning. Did Nichols know he was harboring the Blue Parka Man? And if so, why was he doing it? Or, for that matter, who might be whose prisoner at the almost deserted roadhouse?

The evening passed quickly. The candles, usually snuffed out at nine o'clock, still glowed brightly at midnight as Hendrickson's droll stories and sleight-of-hand tricks kept the cabin rocking with laughter.

Here was a man, Tom figured, who could enjoy everything in life, from a cup of coffee to a woman, with a zest most men never knew, yet who placed no value on life. None of it touched him; the moment had no beginning and led to no end.

Finally, as the candles started to sputter, with the wicks barely standing in their pools of wax, everybody decided to call it a day. Tom tossed an extra blanket to Hendrickson, who sought the only vacant floor space for a bed. A deft roll and he was settled in, snug as a bug except for an arm free and handy to the rifle.

As Tom drifted off into a fitful sleep, he felt that he had the answer to his final question. His meeting with Hendrickson had been no accident. Interspersed between Hendrickson's stories and tricks that night were carefully interjected questions about all aspects of the partners' hunting. The man seemed insatiably curious about every phase of their activity. It was then Tom realized that their every movement had probably been carefully observed, that they were not the only stalkers of the divide. Hendrickson had made the same mistake Tom almost made, watching Tom and waiting too long in rapidly descend-

ing clouds and darkness to be able to find his way back to
the roadhouse.

The following morning the partners were up at the customary early hour. The hunting was good now, and every
one of the lessening daylight hours must be made to
count. Breakfast was a quick affair with none of the preceding evening's leisureliness. Hendrickson pushed back
his chair when he was finished, reached over to pick up
the rifle, his sole possession, and said he would be mosey-
ing along. In the light and with Tom's directions, he said
he would have no problems.

The way people wandered casually into one another's
lives and out again, often to meet years later with the
same casualness, mindless of intervening time, was
strictly a Northland custom. It was like that now. Only
Tom had not seen the last of Hendrickson—not by quite a
lot.

It was almost two weeks later, past mid-October, when
Tom set out disgustedly for the Miller Roadhouse, some
twenty miles north and beyond blizzard-swept Eagle
Summit. For some unaccountable reason Pete Malone
had failed to arrive on schedule, and the rapidly dwindling stocks of flour, coffee, sugar, and feed for the two
horses had to be replenished. The partners had drawn
straws that morning to see who would make the trip. Tom
had lost. He shoved his Winchester into its saddle boot as
the final preparation for departure.

Tom skirted the roadhouse and headed for lower
Twelvemile Creek, slouched in his saddle. The only
sound of man in the miles of wilderness he traveled was
his own frosted breathing and the tiny tinkle of the bell
on the trailing pack horse. Nothing in the monotony of
cold and the snow-blanketed landscape sounded a warning, but suddenly he was acutely aware of danger. He
halted his hand in its first instinctive movement toward

the rifle boot, knowing this was no danger of the wild. He reined in his horse without turning before he heard the voice.

"Where are you headed, Tom?" it asked.

Tom swung at the sound to see Hendrickson stepping from behind a spruce tree that, brushed by his shoulder, cascaded a small avalanche of light, powdery snow over his head and shoulders. A shrug and most of it was gone. Hendrickson's rifle was cradled protectively in his arm, but Tom knew that the instant before it had been trained exactly at the back of his skull.

Nothing burned him like a gun pointed at him. There was only one use for a gun—killing. Fury raged in him. He wished for a chance to draw with Hendrickson, but the latter was grinning, the Savage pointed groundward now.

"I could ask you the same damned question," Tom said.

"Don't get riled up," Hendrickson replied easily. "Nichols gave me a grubstake to sink down some holes here. Haven't seen a color yet. Looks like Nichols won't have much to show for his grubstake."

"Don't look like you have worked too hard at it. Them's mighty puny holes for a man of your size."

"It was a puny grubstake. A man gets what he pays for."

"I'll be seeing you," Tom said, feeling good again and thinking that this was not going to be such a bad trip after all.

"You may," Hendrickson answered.

The winter of 1905–1906 was one of the coldest that veteran gold seekers could remember. It started early with a vengeance, as though the Arctic was determined to fling from its frozen soil the odd assortment of adventurers and gold-grubbing argonauts who had dared penetrate its solitude. The steamer Light, one of the last boats from Dawson to Fairbanks that fall, froze in far short of its destination in mid-October. Its one hundred passengers, dressed in the oddest assortment of winter trail garb ever seen in the North, were forced to walk sixty shivering miles to Chena. Old-timers still speak of that winter as a milestone. All those who wintered through it became the breed of the North. Thereafter the term *chechaco* was reserved for later arrivals.

Tom's trip to Miller House was wasted effort. Alec Miller, the owner, was himself desperate for supplies and was hoping each day for arrival of the Circle City freighters with his winter's outfit. The steamer carrying winter supplies for much of the Circle district, like the steamer *Light*, had been trapped by the early freezeup. It had been forced to "freeze in" for the winter in a slough some twenty miles below Circle.

Alec guessed that the freighters would be arriving with his supplies as soon as they completed hauling the steamer's cargo to Circle. Tom could see no sense in waiting for that indefinite day. Taking what few supplies Alec could spare, more out of friendship than from plenty, Tom turned his tracks back to the hunting grounds. Pete Malone might even now be unloading their supplies at the cabin.

That hope was not to be realized. There was still no sign of Pete when Tom returned. The few supplies he had been able to obtain at Miller House provided slight relief. The partners waited another five days before finally admitting that there was no choice but to head for Fairbanks. It was now Tuesday, October 24. They would have preferred to wait for a break in the intense cold, but it was as unpredictable as Pete's overdue arrival. The decision was made to start at dawn next morning. It is interesting that the thought was never entertained by any of the partners that Fred Nichols's Twelvemile Roadhouse was a possible source of the supplies they so badly needed.

All the partners agreed that this was not one of the easiest trips they ever made. The hours for traveling were short now, daylight having dwindled to eight and a half hours a day. That was normal at this time of year, but not the prolonged and intense cold, which was generally reserved for December and January. The cold slowed their travel.

Two things a man had to know in this temperature: never work up a sweat that, despite absorbent wool clothing, turned to a sheen of ice on your skin when effort slowed. Second, never move so fast that you had to gulp the searing cold into fragile lungs. Thus not until midafternoon of the third day did Tom and his partners top the final low ridge overlooking Fairbanks. No town greeted their view from where they stood, only a heavy blanket of smoke from hundreds of chimneys. In the motionless air that always accompanies deep cold, the smoke rose in unwavering columns to merge in a thick, unmoving curtain. Everywhere else the sun shone in brilliant purity from a flawless blue sky.

The partners had arrived at Cleary too late to catch the stage for Fairbanks, but they were not to be delayed an extra day so near their destination. Hot food, an hour spent in relaxing but almost stifling heat, and they were ready to strike out again. As they stood overlooking the town and the valley, they cheered their decision. The hardships of the journey never seemed as important as the prospects at the end of the trail.

As soon as the partners reached the Modern House, they separated by unspoken consent. No matter how good a partner a man is in the bush or on the trail, he is not what one has in mind for a night on the town. Nothing was said about plans for the future. It was not the time or the place. They were bound to run into one another sooner or later while making the rounds, and that would be soon enough to talk business. Tonight was for pleasure alone.

Tom wasted little time with a bath. Later he could steam out luxuriously at the Finns. He shaved quickly and headed for the Floradora. Here the tinkle of glasses, the clinking of poker chips, and the whir of the roulette wheel was music to a wilderness-starved man. Toward the rear of the hall Tom could see Ham Grease Jimmy calling the dances, his clappers keeping things lively, looking as dapper as he had at the Dawson halls seven years before.

Tom was just raising a glass to his lips when he felt a hand on his shoulder. He knew without turning who it would be.

"Just heard you had reached town," George Dreibelbis said. "Hunting pretty good this fall?"

Tom knew it was a different kind of hunting Dreibelbis had in mind.

"Fair," Tom replied, waiting for the next question, knowing what it would be, and wondering how he would answer it.

"Have you seen anything of Hendrickson this trip?"

There it was, out in the open at last. Why did he hesitate to answer, Tom wondered. George was an old and trusted friend. He did not owe Hendrickson a tip of his hat. It was, he guessed, an old axiom in the North that a man never gave information to the law. After all, there were not many here who could afford to have their pasts examined too carefully. Tom's answer might have been an abrupt "no," but he saw again the cocked rifle in that meeting atop the summit, felt anew the chilling sensation of the encounter on Twelvemile Creek, and with it felt his rage mount as before.

"I guess I did. We met a couple of times," Tom replied.

Dreibelbis waited.

"He's at Twelvemile with Fred Nichols," Tom said at last.

All of a sudden Dreibelbis was wearing the broadest grin Tom had ever seen on the lawman's face. "Not anymore he's not. I just needed some verification he was in that neck of the woods. If it makes you feel better, we had word yesterday that Hendrickson is holed up on Birch Creek. We're setting out to pick him up tomorrow, me and Frank Wiseman."

THE TRAIL

When word reached the marshal's office that Hendrickson was believed to be on Birch Creek, Marshal Perry was spared a decision. Actually, that decision had been made for him on the day of Hendrickson's escape. Now, when he passed on this latest news to Dreibelbis, all he had to say was, "Who would you like for a partner?"

Without a moment's hesitation Dreibelbis answered, "Frank Wiseman."

Marshal Perry raised an eyebrow. Wiseman was the deputy at Cleary City. If this job took too long, he would have to be replaced temporarily by a man from his own office. Perry, however, raised no question and offered no suggestions. All the details were left to Dreibelbis.

Dreibelbis knew there was no better man than Frank to trail with on a case like this. Frank was an old hand at it. He had been a marshal at Haines, the beginning of the Dalton Trail and one of the chief routes to the Klondike at the height of the wild Dawson stampede. When that port city had needed a strong hand, it had been found in Frank Wiseman.

With this most vital of all arrangements settled,

Dreibelbis turned his thoughts to the myriad details requiring his attention. After notifying Wiseman of the plan, he visited the dog barn where he boarded his team and played briefly with each yowling, pawing, frantic animal in turn.

"Take it easy, you fellows. Save all that energy for tomorrow. You'll need it. We have a long hike ahead."

The noise continued unabated until, at a low command, each dog dropped to its stomach, silent, unmoving except for the occasional flip of a tail.

Dreibelbis gave detailed instructions to the gnarled old Indian helper how each dog was to be fed that night, the amount of dried salmon, tallow, and rice to be set aside for the trip, and warned him not to forget the leather moccasins the dogs might need for their feet. Long before Dreibelbis met Tom Gibson that evening, all the food for the trip and his personal gear had been assembled, stored in waterproof duffel bags, and delivered to the dog barn, his rifle and six-gun newly cleaned and oiled.

Actually, the meeting Dreibelbis engineered with Tom was not a necessity. The tip from Eagle Creek Roadhouse, for such was its source, had appeared unusually reliable, but Dreibelbis valued Tom's observations. It never hurt to double check, and he was glad that he had. Tom answered one important question that had been bothering him. Hendrickson was supposed to have reached Eagle Creek Roadhouse only a few days before, so he would have had to shack up somewhere else for quite a spell. Fred Nichols! That surely was an interesting and not altogether surprising bit of news.

Only one more thing needed attention. Dreibelbis headed toward the dog barn where all of his outfit had been assembled. One thing he never allowed anyone to do for him was to load his sled. He now proceeded to do this, making certain he would know where every vital piece of equipment was stored, certain too that when he picked up Frank and his gear next morning it would require a minimum of load shuffling and rearrangement. He then laid out all the dog harness, checking again every snap, collar, buckle, and tug line. Every dog watched intently, hopefully, but only an occasional low whimper was to be heard.

When Dreibelbis emerged from the hotel the following morning, he spat into the air. There was a distinct crackling sound as his spit hit the frigid air. Like other old-timers of the North, Dreibelbis knew that temperature could be gauged by the way spit froze. When it crackled on hitting the snow it meant fifty below zero, but if it crackled in the air it was at least sixty below. Damn, how long could this cold spell last? Even daylight would bring slight warming at this time of year when the sun finally raised its sleepy head.

Dreibelbis was dressed to take it. Long ago he had learned to dress for the trail in the style of the Indians: fur mukluks extending halfway up the thigh over fur pants worn with the fur turned inside; fur parka falling just above the knees, its hood trimmed with a wolverine ruff; a marten-skin hat. There was no beating fur, for it weighed a fraction of white man's dress while providing more warmth, and its loose fit allowed completely free, unrestricted movement.

The natives had developed the ideal winter garments for this northern climate, and the white man was quick to recognize and adopt them. In contrast, the Indians, who were moving into town in increasing numbers, were abandoning their furs for the white man's most nondescript dress, picking up once again the worst of the white man's ways.

Dreibelbis had little time to philosophize that morning, for his thoughts were occupied with plans for the day ahead. After almost three months of waiting, he was elated at the prospect of action at last.

He would be following the same trail Tom and his partners had traveled as far as Twelvemile Roadhouse, just beyond the high summit crossing with its vicious, changeable moods. From there on he would still be following in Tom's footsteps on the latter's recent trip to Miller House, but only as far as the roadhouse that stood at the mouth of Eagle Creek. Fortunately he and Wiseman would not have to tackle Eagle Summit.

Eagle Creek stemmed on the flanks of the summit. It was one of the shortest and richest of the several creeks that joined to form an unnamed tributary of Birch Creek. And it was somewhere on the stretches of Birch that Dreibelbis expected to find Hendrickson. What had taken him there? There were no mines on Birch Creek, no habitations of any kind. He was doing his trackers a favor, for on the unbroken snow of Birch Creek, fresh tracks would leave a trail easy for a blind man to follow.

On arrival at the dog barn, Dreibelbis's first act was to make sure that he had not neglected to pack his thermometer. It was a unique instrument that Dreibelbis and all old-timers had learned to carry on the trail. The store-bought variety was useless at forty below, for the mercury simply dropped to the bottom of the tube and shivered there in a tight little ball. The stampeder's thermometer consisted of three small bottles filled respectively with kerosene, painkiller, and 100-proof Hudson Bay rum. The kerosene congealed at fifty below, the painkiller at seventy below, and the Hudson Bay rum—well, if it congealed, just forget the whole thing and drink up. It never actually worked quite like that because the rum never did freeze, but if it began to thicken it was a sure sign to make camp in a hurry.

Dreibelbis needed no help from the droopy-eyed watchman at the barn who, it appeared, could use considerable assistance himself after a somewhat less than sober night. In the flickering, smoking light of the oil

lanterns, the team was quickly harnessed. The watchman was able to open the barn door, barely wide enough for the team and sled to squeeze through. Even so, the cold, sixty-below-zero air rushing into the barn and meeting the warmer air created a dense, swirling fog.

"Hike!" called Dreibelbis to the lead dog. The team surged into motion, whipping through the door out into the pitch-black predawn darkness. It was 4:30 A.M., Saturday, October 28. The first thin shading of light was yet to touch the unseen horizon.

The air was motionless in the intense cold, but the wind created by the team's swift pace brought quick tears to Dreibelbis's eyes that froze instantly on unprotected lashes. Sight was impossible until he could pull his protecting parka hood up over his head and gradually work the ice free. Bingo, his leader, would take care of things until then. The trail-wise dog knew his job without orders. Soon he would be slowing the younger, impetuous dogs behind him to a more sober working pace. Bingo had watched intently as the sled was loaded with the camp stove, snowshoes, extra clothes, grub, and bedrolls. It all added up to a tough trip and heavy pulling for six dogs.

Dreibelbis's thoughts were on the trip ahead. He was glad that he would be picking up Deputy Frank Wiseman at Cleary. Two men were a lot better than one on the winter trail, where anything could happen.

The trail he followed now was well packed and fast from the constant travel between the mines and Fairbanks. It would be different beyond Cleary. The team was still hitting a fast pace at Pedro Creek, sixteen miles from their start. Dreibelbis found himself wishing that the dogs would slow down, give him a chance to run a little and get the circulation moving in his feet again. He could feel his toes beginning to sting despite the fur mukluks, wool insoles, and three pairs of heavy siwash socks. Well, the dogs must slow soon enough now. Ahead, the trail rose swiftly to Cleary Summit, an elevation of 2,300 feet. They had started from a 400-foot elevation at Fairbanks.

The five-mile climb from Pedro Creek to Cleary Summit would have killed the average clerk in any of the well-heated, comfortable stores of Fairbanks as he peered through frosted windows at the unmoving mercury in the town thermometers. Trail men were not so concerned about exact degrees or elevations.

To Dreibelbis and men of his kind—the dog-team mail carriers, prospectors, and hunters—the Cleary Summit trail was simply the tedious part of a normal day's work. At least he could keep warm now. He shouted encouragement to the dogs as they bucked harder into their collars. He dropped off the sled runners to jog along behind the team, careful to keep his mouth shut lest the paralyzing cold sear his lungs. When he could run no further without gasping for breath, he would jump on the sled runners again, kicking with long, even, powerful strokes to ease the pull on the dogs and to keep them moving. At the foot of the steeper ascents, he would stop the team briefly to breathe the dogs. Then they would be off again, the man and his dogs, all appearing pygmy-size against the huge expanse of mountain and the limitless barrier of snow—had there been anyone in this solitude of time and space to bear witness.

The light of day was just beginning to touch the world to life when Dreibelbis reached the summit of the Cleary Divide. Saffron hues bathed the entire landscape with indescribable beauty. The view across the wide-gouged valley of the Chatanika to the distant sentinel-like White Mountains was enough to make any man pause, even on such a day as this. Here on the summit it was several degrees warmer than on the valley floors where the heavier, colder air hung motionless, but it was no thanks to the sun, which suspended like a copper disk, provided as little warmth as light. Except for an hour or two at its height, a man could look it straight in the eye and never blink.

A barely discernible ripple of air stirred here on the summit, but it had an edge as rough as a file. Perhaps it was the salvation of man in this alien land that he was kept too busy to wonder at its contradictions. Swiftly Dreibelbis pulled rough locks from his trail bag and strapped them to the sled runners. The descent in places was almost perpendicular, gouged with the deep-grooved brake marks of many previous descents. Dreibelbis rode the brake hard to keep the heavily loaded sled from over-running the straining dogs. His leg grew numb from the constant pressure on the brake until the trail leveled briefly and he could switch to the other leg. The trail was icier than usual. Dreibelbis was thankful for the extreme cold, which kept the game from moving, and for the well-trained Bingo at the head of his team. The unexpected sight of moose or caribou around a sharp bend on such hazardous descents as this had often set dog teams off in wild pursuit, occasionally with fatal results for both team and driver.

The spray of snow kicked up by the gouging brake followed the sled like an icy streamer and soon caked Dreibelbis's parka hood, back, and shoulders with a thick layer of snow. He had timed his departure in order to reach the summit in the weak light of day, and he was thankful now for the light that aided him in this treacherous descent. It required but a fraction of the time consumed by the laborious climb up the south slope, yet it seemed infinitely longer until the trail leveled off a final time for the remaining easy miles to Cleary City.

Dreibelbis could see Frank's gear already stacked by the side of the street as he neared the cabin. As he braked the sled, light flickered off inside the cabin. Even as the dogs slowed, Frank pushed through the door. Bundled in his trail furs, like Dreibelbis, he looked, despite his skinny frame, broader than the doorway.

"Straight as an Indian" might have been an expression coined to describe Frank Wiseman, only Dreibelbis had never seen any Indian who carried himself that erectly. His carriage and spare build added to the impression of height; although he stood only an inch above Dreibelbis, he seemed even taller. His movement, like his speech, appeared always calculated, as though he never acted from impulse. Nor was he ever accused of a sense of humor. Dreibelbis, however, had not selected Wiseman to be the life of the party. He knew of no other man so dependable. Frank was as erect in his moral code as in his bearing.

"Where'd you find that scrawny, moth-eaten bunch of dogs?" Frank asked, knowing this team was Dreibelbis's pride and joy. "Probably all of 'em will die on the trail."

"They will freeze you to death before we get back," Dreibelbis answered. "Get your gear aboard. You're wasting time."

Shortly beyond Cleary City the trail picked up the broad, frozen surface of the Chatanika River. Miraculously, despite the intense, prolonged cold, the dread overflow was absent at the start.

"I don't understand it," Frank said, puzzled. "The upper streams should be frozen right down to the gravel by this time and water flowing six inches deep over the Chatanika. It's the worst stream for overflow in the country."

"Relax and enjoy it," George replied. "Bingo can smell open water a mile away. He'll warn us."

Indeed it was fortunate that Bingo could and that he did. They had been following the twists and turns of the river for nearly two hours when Bingo began to whine softly and to slow the team in its mile-eating gait.

Dreibelbis realized that he had become mesmerized by the steady pace of the dogs, the soft crunch of snow beneath the sled runners, and the changeless unfolding of one bend after another on the river. Now, suddenly, he was acutely aware of impending danger, and he damned his carelessness that had left sole responsibility to his lead dog.

Bingo brought the team to a stop even as Dreibelbis barked out the command to halt. The snow felt spongy, and as Dreibelbis lifted one foot, he saw a distinct settling of the snow and a talltale seeping of moisture into the depression.

"Haw," Dreibelbis yelled at Bingo. But though the leader turned his head to the left in acknowledgement of the command, he did not move.

"Gee, then. Go gee, Bingo." If only they could reach the safety of the riverbank, so near yet hopelessly distant. Again Bingo refused to move. Dreibelbis realized that the leader must have sensed the overflow to be deeper beneath the concealing snow along the right and left banks and that the center of the river, slightly humped, was their only chance of escape.

Dreibelbis deliberately steadied his voice. "All right, boy, come back, Bingo, easy, easy does it."

Bingo acknowledged the order with a barely perceptible twitch of his tail and began a tight, slow turn to the right. When the eager swing dogs started to move too swiftly, Bingo snarled and kept them in line. Bingo directed the maneuver, making sure each dog buckled into his harness and turned precisely on cue. Dreibelbis, applying all the pressure he could on the handlebars and side of the basket, but with the main help coming from the dogs, gradually turned the sled until it faced downstream again.

The entire operation only took minutes. Yet the water

rose so swiftly that by the time the team was again in motion downstream, water had reached almost to the upper sled stanchions. Dreibelbis's feet were soaked to the ankles. The once slick sled runners, polished with hours of patient care, ice-covered now, bumped rather than slid across the snow, the sled bouncing clumsily. By the time the safety of a gradually sloping bank was reached some five hundred feet downstream, Dreibelbis had lost all sensation of feeling in his feet.

Frank hit the snow seemingly an instant before the sled touched the bank. A final desperate lunge by the team with both men pushing and the sled was out of danger. Dreibelbis felt himself pushing into a sitting position against an overhanging snag, his feet deeply buried in loose snow as absorbent as cotton. Almost immediately Frank had a fire crackling nearby. Not one word had been exchanged, and only now did Frank ask Dreibelbis where his extra footgear was stashed in the load. With swift strokes of his hunting knife, Frank cut the stiff covering from Dreibelbis's feet until they were bare, stretched out to the warming blaze, and then began to massage circulation back into the whitened toes. Soon fiery, stabbing pain set in as feeling was restored. Dreibelbis winced in pain.

"That's what we want," Frank said, kneading the feet a bit longer before tossing the fresh footgear to Dreibelbis. "They will be tender for a few days, may even raise a few blisters. Sure you want to go on?" Frank asked.

"You're damned right I'm sure," Dreibelbis answered as he struggled painfully to get his feet into fresh socks and mukluks. He glanced up to see Frank slip to his knees, his arms encircling Bingo's neck in a strong, deep hug.

Dreibelbis pretended not to notice. Bingo, accustomed to affection only from his master, pulled away slightly and, like the other dogs, continued to chew at the ice balls that had formed between his toes. Without either of them ever mentioning it, both Dreibelbis and Wiseman knew that this was one they owed strictly to Bingo.

Frank rose to his feet. "Guess this is as good a time as any for lunch," he said. Dreibelbis allowed that it was. Snow was quickly scooped into a coffeepot and set to heating over the fire. Coffee was added and allowed to come to a strong boil. The indispensable coffee, cold bannock, and some strips of squaw candy—the dried salmon cured for human use, and favored on the trail because it required no time to prepare—provided an ample lunch. The dogs' feet were checked a final time to make sure all ice was removed and, unknown in the lexicon of all accepted dog training, each dog was tossed a strip of salmon. It had been their show.

It was slower going now off the river, in loose snow and a mostly unbroken trail. As the afternoon wore on, Dreibelbis and Wiseman traded places on the runners, and occasionally one or the other took turns running behind the sled. Fortunately the snowfall had been light this winter and there was no need to snowshoe ahead of the team. They were making good time, if only Dreibelbis's feet would hold out. Running behind the sled was becoming increasingly painful, his stride an awkward hobble.

This was dead country they traveled now, the only sound the steady creak of their sled runners. The game had holed up with the onslaught of the cold. The impermanence of man had been observable throughout the day in the abandoned and fast-decaying cabins of this once bustling trail. Except for these relics of a recent past and the gnarled, twisted spruce, there was nothing to break the endless monotony of glaring, unrelenting whiteness.

For the last hour through deepening shadows the valley had narrowed rapidly as though threatening to snuff out all avenue of escape. Then, in the last fading light, a

cabin loomed up ghostlike on the edge of the high riverbank. George broke trail through the powdery snow up a
sharp incline while Frank rode the brake to slow down
the clamoring, eager dogs who sensed the end of the day's
trail and their night's feed.

The two lawmen had chosen this stopping place by
mutual consent, knowing it lay in one of the few sheltered recesses in the hills and that the deserted cabin
boasted the only stove on this part of the trail.

Dreibelbis set to work at once preparing the dog food, a
rich mixture of cooked rice liberally thickened with dried
dog salmon and chunks of tallow. It cooled rapidly when
set outdoors in this temperature and was soon ready to
feed. The entire process had been accompanied by an
incessant howling and barking that reached a crescendo
when Dreibelbis ladled out the first helping to Bingo.
Ordinarily he never permitted such noisy demonstra-
tions, but feeding time was an exception, and today the
team had earned the privilege many times over.

Meanwhile Frank had been equally busy with the lawmen's dinner, and it was just as utilitarian as the dogs'
meal. Trailmen invariably kept a huge batch of red beans
liberally laced with bacon already cooked, frozen, sacked,
and ready for the trail at a moment's notice. Now all
Wiseman had to do was hack off enough of the beans for
the night's meal, set them to thawing and bubbling in a
greased pan, bake enough bannock to go along with it and
for the noonday lunch next day, and set the inevitable pot
of coffee to brewing. This culinary banquet was complete
at just the moment Dreibelbis was feeding the last of the
dogs.

At last, with the dogs satisfied and bedded down, their
own dinner finished, and shed of their trail clothes, the
lawmen were free to relax for the first time that day. The
roaring fire turned the sides of the rusted, potbellied
heater a cherry red. After the bitter cold of the trail, they

could feel drowsiness already taking hold. They were not talkative men, and the subject of the Blue Parka Man had barely been discussed. When they did meet up with Hendrickson, previous talk would play no part in the outcome. It was sufficient that each man placed complete confidence in his partner to act as the situation demanded. Dreibelbis knew he and Frank had been fortunate that afternoon to escape with their lives. He realized how lucky he was to be sitting by this fire, enjoying its crackling warmth. It would have been a different story had he been alone. Yet it was this very aloneness that Hendrickson had faced and survived these past months. The day's experience bred renewed respect and growing caution for a foe who had lived so long with danger.

Dreibelbis shook his head to break the spell and rose painfully from the wobbly box he had been using for a chair. "I'm bushed," he said. He hobbled toward the door and gathered up a parka. "Check the dogs and I'm ready for the sack."

Frank waited until the cold draft from the quickly opened door had dissipated somewhat, then hurriedly stuffed another chunk of wood into the stove, stripped to his long johns, and dove for his bedroll. But he was not quite ready for sleep, not until he would hear the door open and know that all was well. But it didn't happen. Frank stirred uncomfortably.

With sound muffled by the heavy plank door, Frank barely distinguished George's words. "God man, get out here. The lights, man, the lights."

Frank scrambled out of his bag mindless of the stinging cold on feet still protected by heavy socks. He flung on clothes and boots, wondering if George had suddenly gone loco. He knew it could happen just that suddenly and inexplicably. He threw open the door.

There was George, standing transfixed, his gaze riveted on the heavens. And then he saw them too, the Northern

Lights as he had once heard them described. He had attributed that description to drunken imagination. Everyone had seen Northern Lights, but seldom such as these. They streamed across heaven in a thousand constantly changing shapes, rippling with all the colors of the rainbow, so low he knew he could have jumped and touched them. The heavens were so aglow with color that not a star was visible. The constantly wavering, changing shapes were accompanied by a crackling, explosive sound as though God were cracking a whip to make the streamers dance.

They stood unmoving, marveling. They might have turned to stone, or more likely to columns of ice, had not the spell been broken by the whimpering of the dogs. Shaken, uncomprehending, they turned to the only god they knew, and man had to assure them of his invincibility. Kind hands ruffled their fur, stroked their heads, and reassured.

Unspeaking, the men returned to the cabin. The lights were never discussed between them until, many days later, they tried to describe this sight and its accompanying sound to their fellow deputies. Amused glances and sly winks greeted their account. It was enough that they had seen and heard.

Before reaching Twelvemile Roadhouse at the end of the second day's travel, the trail would rise from the Chatanika Valley floor some 1,300 feet to the soaring 3,000-foot crest of Twelvemile Summit, second only in height to Eagle Summit in this southerly spur of the White Mountains. This was the divide between the Tanana and Yukon river drainages, and it derived its name from the fact that twelve miles to the northeast lay the rich placers of the Birch Creek district: Mastodon, Mammoth, Independence, and Eagle creeks among them.

The second day started as the first had ended, in bitter

cold that seemed to know no end. Trees popped and split in the otherwise awesome stillness as their fiber contracted and separated. Neither Dreibelbis nor Wiseman could be certain when they detected the signs of change. Perhaps it was so subtle as to be indistinguishable at first, but more likely neither man would admit to a hope that the cold had started to break. Dreibelbis was on the sled runners when the realization dawned. He braked the dogs to a stop, threw back his parka hood, and inhaled tentatively, then deeply. There was the fragrance of life in the air again.

Rarely are the warm chinook winds of the coast able to break through the mountain barriers to reach the interior. Rarer still do they warm the temperature so rapidly. Now men and dogs were enjoying a second miracle, but this one within the realm of their understanding.

Their celebration over, Dreibelbis yelled at Bingo to hit the trail again. Until the noon stop, two hours later, they maintained a steady pace. Parkas were shed as the temperature soared to thawing. Alongside the trail the warmth and the breeze combined to send the snow cascading from the spruces, and the men could smell their resinous fragrance. This was travel at its best.

7

TWELVEMILE SUMMIT

From a slight, barely discernible breeze, the wind slowly increased in strength. It grew in warmth as it gained intensity, blowing at their backs and aiding their progress. It was not until the two men rounded a bend in the trail and the wind hit them from another direction that they became aware of its growing power. It was no longer the gentle breeze of midmorning but gave promise of developing into a genuine howler. From relief at the end of the cold spell, the men's mood changed to mild concern, then to outright anxiety. When they stopped for their noon lunch, it was in a sheltered niche in the hills with Twelvemile Summit towering and indistinct above them, at times completely obscured by a whirling frenzy of driven snow.

Neither Dreibelbis nor Wiseman wanted to utter the words first. Finally they came tentatively from Wiseman's lips.

"You suppose we ought to camp here and wait her out?" he asked.

Dreibelbis's reply was slow in coming as he weighed all possibilities. Good sense dictated that they pitch camp. They had a tent, all of the gear necessary for adequate shelter. If not comfortable, they could at least be safe here until the storm blew itself out. But how long would that take? Dreibelbis knew that such a storm could continue for days and could possibly worsen. If that should happen, their food would never last. Dreibelbis had figured on camping out for two nights at most. For the rest they could count on Twelvemile Roadhouse and the roadhouse of sorts that Sid Wilson operated at the mouth of Eagle Creek. Their food might be stretched for another two days at most. Then it would be starvation for men and dogs alike. Hendrickson did not enter into Dreibelbis's calculations at this point, for in a storm such as this he would certainly be pinned down as tightly as they were. They could always catch up with him. It boiled down to a choice between safety—if there was ever such a thing—and possible starvation. It was not much of a choice, but it had to be made.

"I opt for the summit," Dreibelbis said at last.

That decision suited Wiseman. For men of their stamp it was a logical choice. Action was always preferable to waiting.

Dreibelbis took the sled runners and yelled to Bingo. As they left the protected gully of their noon stop, the intensifying storm met them head-on. The team came to a near halt in the face of the blast. Dreibelbis shouted to Bingo again. The howling wind tore the words from his lips and hurled them unheard toward the desolate, wind-swept summit. Bingo, half-turned awaiting a command, never heard the words but knew what George wanted. Some of the dogs behind him whimpered as the wind clogged their eyes, ears, and nostrils with wind-whipped snow. Bingo hackled to the tip of his tail, fur standing in a spiked ridge down his spine, at this display of cowardice. Because they dreaded Bingo worse than the wind, the unwilling dogs faced into the teeth of the storm.

There was no stopping now—no more timber and no safety short of the valley beyond the summit where the roadhouse nestled. The wind whipped at the tightly lashed load trying to tear it apart. Failing in this, it tossed the heavily loaded sled as though on the crest of a wave. The dogs bellied down, scratching for precious inches, the hard-packed snow as polished as a dance floor under the ceaseless pounding of the wind.

Dreibelbis could barely distinguish his wheel dogs hitched nearest to the sled. The rest of the team, Bingo leading, was lost in the drifting snow. Time lost all meaning, being measured not in minutes or hours but in hard-gained yards. Enveloped in the driving snow, men and dogs inched forward in near total darkness, worming their way past precipitous chasms and over ice-encrusted ridges. No man alone could survive the fury of the wind. It was Bingo's unerring judgment that brought them through, judgment born of a close man-and-dog partnership that wrote countless epics into the conquest of the North.

Finally the wind began to lose some of its fury and the sky lightened to reveal the dim glow of fading daylight. They were over the summit and in the lee of a protecting shoulder that broke the main force of the wind. But they dared not stop until the protection of timber was reached.

Once in the spruces they built a roaring fire and turned the team loose to help the near-blind dogs melt the ice from eyes and nostrils. Finally they brewed a pot of coffee that almost stood alone. Dogs and men were stiff and sore from the pounding of the wind. They moved clumsily and off balance, each movement an aching pain. But it was only a few more miles to Twelvemile Roadhouse, where a warm barn awaited the dogs. And there, too, would come the long-awaited confrontation with Fred Nichols. Bingo sensed his master's urgency in those last few miles and hauled the stiff, aching team into a fast trot.

In sight of the welcome roadhouse, Frank hurled himself from the sled and raced to the barn. He had the doors open before the weary team reached it. Without pausing, Dreibelbis drove the team into warmth and comfort redolent with the scent of the summer's high prairie hay.

He moved stiffly from dog to dog, patting each head. They were still too weary to even roll, but each tail thumped gratefully.

"Get on with your business," Frank said. "I'll bed down the dogs."

With a final pat for Bingo, Dreibelbis headed away. As he opened the barn door, he paused to glance behind him. There was Frank cross-legged on the floor with Bingo already in his lap, Frank massaging and kneading the tired muscles.

On the way to the roadhouse George met an Indian helper in hastily donned parka and slippers, cursing the cold and obviously disgruntled at this unexpected arrival.

"Nichols at the roadhouse?" George demanded.

An affirmative grunt was the only reply.

"Put the dog pot on to boil," George ordered. "The dogs had a hard trip and need a big feed. And see that you do it right or I'll feed you to the dogs."

Dreibelbis knew that the warning was unnecessary. After today's trip Frank would skin the fellow alive if the dog food was not cooked properly. Frank was a stickler for details.

It was characteristic of Fred Nichols that he had never appeared to welcome the travelers. Any other roadhouse owner Dreibelbis knew would have met whoever traveled this day's trail with solicitous offers to help. Concern for his fellow man had long ago made a brotherhood of all who traveled Alaska's wilderness. But apparently that word had never reached Fred Nichols.

As Dreibelbis pushed open the roadhouse door, the rush of overheated air was like a furnace blast. Quickly he started to peel out of his trail garb. He was just pulling the parka over his head when Nichols belatedly appeared.

"I suppose you will be wanting supper," Nichols said.

In later times Dreibelbis was to think that this was just Nichols's way of speaking, that the words held no special malice. But they were certainly poorly chosen for this moment.

"You're damned right we want to eat," Dreibelbis shouted, taking a step toward Nichols as he gave the parka a final jerk over his head. "We want to eat. We want to eat fast. And it had damned well better be hot and the best meal you know how to cook." Then, for special emphasis, "And lots of it."

Nichols retreated to the kitchen.

Whatever his other faults, Nichols was a good cook, and he spared no pains in preparing the meal. As soon as his guests finished a dish it was heaped full again. Throughout the meal neither Dreibelbis nor Wiseman spoke a word to Nichols. The food received their undivided attention. Nichols was left to sweat, and he did, mopping frequently at his perspiring face and neck. He had soon recognized his unexpected guests, and their manner did nothing for his peace of mind.

Dreibelbis finally pushed himself back from the table and turned to Nichols, who was hovering nearby.

"Sit down, Fred," Dreibelbis offered.

"I'd better clean things up now," Nichols said.

"You had better just sit."

Nichols sat.

"We hear you had a visitor," Dreibelbis continued after a pause.

"Not many anymore these days. You know, a few now and then."

"Then you shouldn't find it hard to remember this particular fellow. Seems he arrived here in pretty bad shape, starved, and so ragged you even outfitted him with some of your own duds."

How had they learned all this? Nichols was not too worried yet. He had rehearsed his story so often that he had it letter perfect. Still he dabbed at his face with an already damp bandanna.

"Oh, you mean that fellow Johnson. Bad. Bad. Lost all his outfit. Near done in when he got here allright. Spark from his campfire set fire to his tarp. Lost everything but the clothes on his back. Getting along late too, it was. He was near frozen."

Nichols had paused at the end of each statement for dramatic and convincing effect. Now Wiseman rose from his chair to prowl noiselessly about the room. Nichols tried vainly to keep both men in his line of vision.

Dreibelbis let Nichols's words trail off. After a pause he said, "Try harder, Fred. You can do better than that."

"What do you mean, try harder? That is the honest truth, so help me. Think I'd try to fool you fellows?"

"We think," said Dreibelbis.

"Ever hear of the Blue Parka Man?" Frank asked.

"Blue Parka Man?" Nichols stammered. "You aren't saying Johnson was the Blue Parka Man?" He had turned to face Wiseman. Dreibelbis tapped his knee and Nichols swung back, terrified at this double-barreled attack.

"You had better come clean," Dreibelbis said. "Make it easy on yourself. I saw you the day of the trial. You had a grandstand seat. I've already got you for harboring an escaped convict. Maybe I could add to that aiding and abetting his escape."

As a matter fact, Nichols had been nowhere near the courtroom on the day of the trial, but he was now reduced to such abject terror that he was ready to admit to almost anything but to being the Blue Parka Man himself.

"Honest, fellows, I never laid eyes on him until he stumbled in here. You got to believe that." Nichols's words were gushing out in a torrent now. "He was near done in and gaunt as a starved wolf. I never saw the like before, but that rifle, that goddamned rifle, he kept it on me, fingering the trigger all the while. Staggering weak he was, but he could pull that trigger. You know he could do that."

"When did he show up here?" Dreibelbis asked mildly.

"I don't rightly recollect. Not too long ago."

Wiseman in his prowling about the room had stopped beside the window and busied himself scraping away at the thickly iced pane. He returned now to a position behind Nichols and placed his icy finger squarely in the middle of Nichols's neck. It might as well have been the Blue Parka Man's rifle barrel. Nichols collapsed in trembling fear.

"When?" Wiseman asked.

"The thirteenth of September. So help me, that damned unlucky thirteenth. What more do you fellows want?"

"Try the truth for a change," Wiseman said. "That always helps."

"So then he told you he was the Blue Parka Man," Dreibelbis continued.

"No, no. He said his name was Johnson, but it didn't take much figuring to guess who he really was. Who else could it have been? Who is going to go wandering around these godforsaken mountains for no reason?"

"So," Dreibelbis continued, "Hendrickson was here for a month, give or take a couple of days. You fed him, you gave him clothes, you made no effort to get word to us. You must have been fond of his company."

"I was scared I'd be murdered. Sure I helped him. So would you." Nichols hesitated, as though wishing he could retract the last words. "You don't know what it was like cooped up here with that fellow, never knowing what he might do next. I finally paid him to move on."

"You don't say?" Dreibelbis said. "Does seem you could have paid him off sooner and saved your peace of mind."

Nichols had exhausted his store of explanations. He sat limp and silent.

"You know, Fred," Dreibelbis said, rising to his feet, "we will be going over to Birch Creek tomorrow to pick up Hendrickson. We should be back by late afternoon and will be needing meals and bunks for three. Be spending the night with you."

Dreibelbis stretched, yawned, and turned toward Wiseman. "I'll be checking the dogs and then hitting the sack. Looks like you can use some shut-eye, too." He strode toward the door and, with his hand on the latch, turned back to face Nichols. "It will be interesting to compare stories tomorrow night when we have you and Hendrickson together. Should be quite a reunion."

8

BIRCH CREEK

Breakfast was a hurried affair. Although it was not mentioned between them, the lawmen's thoughts were on the hours ahead and what might await them on the lonely stretches of Birch Creek. Whatever that might be, for better or for worse, they wanted it over with as quickly as possible.

As Dreibelbis hitched up his team, he wished he could spare the dogs this trip. They were still stiff from the buffeting of the day before. Today, however, the wind could be their ally. Although the worst of the wind's fury bypassed the valley floor, it screeched with intensified frenzy behind them atop Twelvemile Summit and ahead across towering Eagle Summit, the last remaining barrier to the wide valley of the Yukon. While the wind persisted in its present fury, no one could escape from the valleys between the summits.

Sid Wilson's Roadhouse was strictly a mining camp, its principal building a combination bunk and cookhouse that housed the crew on this, one of the richest placers of the Birch Creek district. Its location on Eagle Creek astride the Circle City trail made it a natural stopping place for travelers and, because everyone was welcome, the word *roadhouse* came to be applied to it. This was the lawmen's immediate destination.

Dawn was just beginning to cast a dim glow when Dreibelbis braked the team to a halt at the long, low cabin, to be greeted moments later by Sid Wilson. Their arrival was obviously as welcome as unexpected.

"Don't know how in God's name you made it in this storm," Sid announced as soon as the lawmen were seated at one of the rough plank dining tables, steaming mugs of coffee and fresh-baked cookies before them.

"We notified the marshals at Circle, too, and haven't seen a sign of them. Don't know how you ever beat them here."

They had been joined at the table by three other men whom Dreibelbis recognized as miners on adjacent creeks. Sid voiced their common concern.

"We will have some big cleanups waiting the spring breakup. Mighty tempting for some easy-conscienced fellow." It was a possibility that had not escaped Dreibelbis.

It was soon apparent that the man they were hunting was indeed Hendrickson. Any doubts to the contrary were dispelled, although no one would admit to recognizing Hendrickson first. Dreibelbis could not blame them. Hendrickson's easy escape the past August could have a way of happening again, they figured, to say nothing of the fact that Hendrickson was still at large. His apprehension was a job they were perfectly willing to leave in the hands of the law, with as little implication attached to themselves as possible.

"He came in here asking for a month's job," Sid continued. "I had this job cutting lagging, that had gone begging for months. Most of my ground is thawed," he said, by way of explanation, "and all the underground tunnels need timber supports. I thought I was damned lucky when he took the job." Sid paused. "Of course, I didn't know then that this fellow was the Blue Parka Man.''

Now Dreibelbis had the answer to Hendrickson's presence on Birch Creek. He knew that the dwarfed spruce growing on these high creeks was not big enough to make an axe handle. The nearest decent timber was to be found only on Birch Creek.

"Most fellows nowadays don't want a job that takes 'em twenty feet from camp," Sid complained. "Hard to find a fellow willing to face the bush alone."

Dreibelbis figured they had learned all there was to know. Time was wasting. He and Frank began struggling into their trail gear, thankfully discarded in the welcome cabin warmth.

"Don't suppose you can tell us where on Birch we will find Hendrickson?" Dreibelbis asked.

"Not rightly. The trail would be plain except for this blasted wind. Just follow down the creek to where it meets Birch. They are cutting upstream some few miles."

Frank, taking a last gulp of coffee, emitted a choking gasp, then his mug banged the table. Dreibelbis's own response must have been equally startled, for he gazed at Sid in momentary dumbfoundment. "Did you say they?" he managed finally.

"I guess I forgot to mention it. Sure, Hendrickson has this fellow Jenkins with him. A real sponger that one. Moved in here last fall and hadn't done a lick since. I finally told him to get out there with Hendrickson or just get out. Didn't sit too well with Hendrickson, but he agreed."

"What else have you forgotten to tell me?" Dreibelbis demanded ominously.

Sid backed a step. "It plumb slipped my mind, that's all. Nothing more to tell you. They was back ten days ago to pick up more supplies. That was when we decided it was Hendrickson for sure."

Almost as an afterthought, Sid added, "I gave them a four-dog team. Good timber is scarce on Birch. They have to move camp frequent."

Watching, Dreibelbis could see Sid slowly realizing that he had, unwittingly, provided Hendrickson with the

Charles Bunnell Collection, Archives, University of Alaska,
Fairbanks

Freighting on the Valdez Trail

perfect means for escape. But perhaps Hendrickson was not yet ready to make his move.

"Let's get going," Dreibelbis called to Frank. Minutes later they were on the sled headed for Birch Creek.

As they left the protection of the small cluster of roadhouse buildings, the wind took a vicious swipe at them, as though determined to tear the breath from their lungs. Their progress was so slowed by the relentless wind that it was an hour before they reached Birch Creek. Dreibelbis braked the weary team to a halt. It was hard to hear one's own voice above the moaning of the wind. But this would be to their advantage, for it would help to muffle the sound of their progress and perhaps add an element of surprise in their favor.

Dreibelbis leaned over the back of the sled and gestured for Frank to raise his ear flaps.

"Let's try to do this without gunplay if we can," he said. Frank nodded agreement. Both men knew what a gaping wound and loss of blood meant in this temperature. Loss of even a little blood froze the life from a man in a hurry.

They resumed traveling at an even slower gait, both men carefully watching the timber along either side of the trail. Twice they observed barely discernible trails of a day or so before, leading back into the timber, almost choked now by the drifted snow. They kept moving, the trail up Birch following the left limit of the creek. Barely half an hour later they found their first fresh sign, a newly broken trail leading out of the timber to the main trail they were following, which continued up the creek.

"Stay with the dogs, Frank. I'd better check this out."

Revolver drawn, Dreibelbis proceeded cautiously up the trail. He had gone but a short distance when he discovered a recently abandoned camp. Ashes dumped from a camp stove were still warm. All suitable lagging material for supporting the sides of the tunnels at the campsite had been cut and stacked. The men had pulled out with all of their gear in search of another good stand of timber. They could be only a few bends of the creek ahead.

Back at the sled Dreibelbis quickly filled Frank in on his discovery. "This might just make it easier," Dreibelbis concluded. "In a permanent camp they would be better set up to welcome visitors."

Frank nodded agreement. They took up the trail again, moving in absolute silence. Bingo, wise to this trailing game, took his commands solely from Dreibelbis's slight pressure on the brake and his accompanying gestures. Fresh sign of Hendrickson was soon evident, the first of several creek crossings into timber on the right limit. The weary nature of this work was easy to read, new trails broken by snowshoes through three to four feet of loose, powdery snow. Leg muscles would be aching after a few hundred feet, the unbroken snow piling high on the webbing, adding weight to be lifted with each plodding step. To Dreibelbis, knowing how this kind of work wore down a man, it was good news. It might just give them that extra edge.

Dreibelbis and Wiseman took turns following each fresh branching, the other staying with the dog team, thankful for the short rest. It was Frank's turn to follow the next trail when, rounding a sharp bend, Dreibelbis spotted a dog team tied alongside the trail not more than a hundred feet ahead.

The strange team immediately set up a clamor that could he heard even above the whistling wind. At the same instant Frank lunged from the sled for the protection of timber, rifle held shoulder high to avoid the snow.

The next moments were like a fast-moving, confused kaleidoscope. About the time Frank reached timber, Dreibelbis dashed past his dogs to the other sled. There, leaning against the sled basket, was as welcome a sight as he had ever beheld: a Savage .30-.30, its freshly oiled stock and barrel gleaming from attentive care. He grinned happily and dropped to one knee behind the scanty protection of the sled, revolver trained.

At that same instant the two men warned by the clamorous dogs burst from the cover of timber a couple of hundred feet upstream from where they had entered. They could not see Frank posted behind a large spruce.

Hendrickson pushed a much smaller man floundering ahead of him through the unbroken snow. Frank waited for them to break into their old trail a little above where he stood, then drew down on the men, ordering them to stop and throw up their hands.

Hendrickson's reaction came with the lithe, ferocious savagery of a wolf. In a whirling, instantaneous movement, he hurled the little man directly toward the rifle, rendering Frank powerless.

Then Dreibelbis shouted from behind him. "Stop or I'll drop you. Up with those hands."

In the moments of his furious drive Hendrickson had seemed to grow even larger in stature, like a charging bear. Now, standing motionless, he appeared to Frank to be reverting to his normal, powerful size. His eyes that had blazed into Frank's, so close that their frozen breaths mingled, still held Frank in a steely glare. Never in those first moments did he acknowledge Dreibelbis's presence at his back by so much as a flicker. Then, ever so slowly, the fury lessened and his hands lifted to shoulder height.

The little man, sensing that he was not the main attraction of this meeting, slipped as quickly as he could from under the rifle muzzle. Furtively his hand darted beneath his parka, reappearing with an object that he hurled quickly into the deep snow beside the trail.

"Keep clear," Frank warned Hendrickson's companion. "And watch what you do." Then, to Hendrickson, "Back up and turn around slow. Make for the sled and keep those hands up."

Slowly the two moved forward until Hendrickson reached the packed trail beside the sled.

"Put your hands behind you easy," Frank commanded. With Dreibelbis's revolver trained squarely at his middle, Hendrickson did as ordered. Frank quickly slipped cuffs over the extended wrists, and with a surge of relief he clicked them shut.

Hendrickson's companion had been a dumbfounded spectator to all these events. Finally he found his voice. "Who are you fellows?"

"U.S. marshals," Dreibelbis answered. "Sure you don't know who you are traveling with?"

The man's bug-eyed astonishment was all the answer they needed.

"I thought this was a stickup." Then, in sudden horror, "My God, my poke. I threw my poke in the snow when I thought we were being robbed." Stumbling, flailing his way back up their path to where he had faced Frank's rifle, he dropped to his knees and began frantically digging through the snow.

Dreibelbis was getting his first look at Hendrickson in nearly three months. The face he searched bore the marks of those months: lean, hard, the telltale signs of many freezings plain. Sweat from the recent exertion of snowshoeing ran out from under the edges of his fur hat and streaked down his cheeks.

Suddenly Hendrickson's face broke into a wide grin. "Like old times, eh, Dreibelbis? But I always seem to be on the wrong side of the gun."

Only then did Dreibelbis lower his revolver. It seemed as though he would have to pry his right hand from the grip, he had been holding it so tightly.

"Made it easy for you this time," Hendrickson said. "I packed that rifle through the drifts all morning, up to now. Don't expect any more favors."

"You can depend on it," Frank promised.

Dreibelbis carefully examined the dogs' feet, then rose, dusting the powdery snow from his knees. "If we hurry we'll reach the roadhouse in time for dinner," he announced.

Hendrickson's laughing, good nature seemed to reassert itself. "I'll get my gear off the sled and run along behind the team for a while," he said. "With all that sweating I'm beginning to freeze."

"Wouldn't bother you for a minute," Frank replied. "I'll get your things. And we want you in plain sight, out there ahead of the dogs."

Frank walked over to the other sled to gather up Hendrickson's personal belongings. Lashed in its scabbard on top of the load, where it could be reached at a second's notice, was a sharp-edged hunting knife. No, Frank thought to himself, he won't be making it easy, and Fairbanks is still three-day's travel away.

Dreibelbis turned the team in a sharp gee back down the trail. Scant attention had been paid to Hendrickson's partner, still burrowing through the snow in search of his poke, like some desperate marmot pursued by an owl.

Now, observing the deputies' final preparations for departure, the man let out a yell. "Hey, you fellows, give me a hand. Least you can do is help me find my poke."

Hendrickson laughed and shouted, "The boys got their hands full with me. You're on your own." He started to lope down the trail ahead of his captors, arms swinging awkwardly behind him in the binding handcuffs.

A torrent of profane abuse followed them until a bend of the trail shut off further sound.

It was only then Dreibelbis realized that an astonishing change had occurred. The voice had carried clearly, not muffled by a screeching wind. He looked quickly at the line of timber; it was no longer twisting and swaying in a wild dance. Then he turned his gaze to the heavens and there, towering high above, clearly visible, was Eagle Summit outlined against the lowering sun. Only a stray wisp of lacy snow occasionally ruffled its surface. Dreibelbis reached up to lower his ear flaps, surprised that his ears were stinging. With the abating wind, the temperature was steadily dropping. He guessed it must be a good twenty degrees colder than when they left the roadhouse that morning. Much had happened since then.

The day crew, just off their ten-hour shift, were filing into the cookhouse when the lawmen and their prisoner arrived. Cautiously, rifle pointed directly at Hendrickson's back, Dreibelbis motioned him through the door, held partly ajar by one of the startled crew. Wiseman followed moments later, having seen to the dogs.

Sid Wilson witnessed the abrupt entrance and hurried forward. Quickly Dreibelbis explained that they needed dinner and wanted the table nearest the wall to themselves. Sid ordered the benches beside it cleared immediately. With scattered grumbling the displaced men crowded into space at the remaining tables, all of them looking for places with the best view of the deputies and Hendrickson. Sid was invited to join the deputies at their table, which he did with noticeable lack of enthusiasm.

While it was unlikely that Hendrickson had confederates among the crew, Dreibelbis was taking no chances. He took his place at the table, back to the wall, facing the room. Hendrickson was ordered to sit facing Dreibelbis, his hands still manacled but in front now so he was able to eat. Wiseman had a place set for himself at the end of the table with both Dreibelbis and Hendrickson directly in view, some five feet away. Sid Wilson, obviously wishing he could be anywhere else, sat beside Dreibelbis. Only when the first heaping dishes of food were placed on the table did Wiseman put down his rifle on the bench on Dreibelbis's side of the table.

The meal progressed swiftly to an accompanying low hum of speculation from the crew at adjoining tables. Most plates were being filled a second or third time when the door burst open. Two men sprang into the room, their rifles drawn, seeming to cover every man in the room. In their trail furs, parka hoods fringed with frost, they seemed of more than human size.

"Keep your hands in plain sight," one ordered. "Don't give us any trouble." They separated slightly and took a further step into the room, the icy blast from the open door behind them sending an almost paralyzing draft through the superheated room.

Wiseman's hand moved involuntarily toward his rifle in that first shocked second of confrontation. Then he stayed it, knowing he would be a fraction slow. In that instant he felt a certainty beyond contradiction that these men were Hendrickson's accomplices.

Now the man slightly in advance of the other threw back his parka hood, quickly brushing the melted frost from his eyelashes. It was then that he recognized Sid sitting chalky white and motionless at the table across the room.

"All right, Sid, just point him out. Leave the rest to us. The rest of you take it easy."

Sid could not find his voice. He tried but no words came.

It was Dreibelbis who broke the silence, and his voice was brittle wih repressed anger. "You can put 'em down, Edgar. You're a trifle late. We've got your man."

For what seemed a frozen eternity there was no movement or recognition. Then, rifle still raised, the man advanced a few tentative steps.

"You George Dreibelbis?" he asked disbelievingly.

"Yes, you damned idiot. Put down that blasted rifle before it goes off."

Suddenly, from deathlike silence a moment before, the room exploded in a bedlam of sound, with everyone talking at once. Benches scraped against the floor and some toppled over as men jumped up, pressing forward for a better view of the proceedings.

The man named Edgar strode over to Dreibelbis's table. He stared hard at Hendrickson.

"So you are him," was all he could say.

For a stunned moment Wiseman could not believe that these men were a deputy marshal and guard from Circle City. He recognized Edgar as quickly as Dreibelbis did, when the man threw back his parka hood, but his mind could not accept the fact immediately. Marshals did not act in this bizarre fashion.

Pale with anger, Wiseman was on his feet before any of the milling, excited crowd could approach Hendrickson. Rifle in hand he fended off the pushing, curious mob, keeping them at a safe distance.

Hendrickson was missing no part of the excitement that swirled about him. In the pandemonium that first reigned, he had jumped to his feet with the obvious intention of disappearing in the crowd, until he saw Wiseman's rifle trained at his belt buckle. With hands raised in mock defense, he sank back on the bench. Better opportunities would come later. In the meantime he could enjoy the scene before him, knowing it would do nothing to ease his captors' nerves. He could afford to wait—but not for too long.

When a measure of calm had been restored, Sid edged over to Dreibelbis's side. "I guess you will want places to sleep for the night?" he asked.

Dreibelbis could not help laughing at Sid's obvious lack of enthusiasm.

"No, Sid, we have to make it to Nichols's place and be

ready for an early morning start to the summit."

All of that was true. Dreibelbis wanted to be across the summit while the calm still held. But Dreibelbis had other reasons for wanting to leave the roadhouse, reasons that made him every bit as anxious to be gone as Sid was to have them gone. There were too many unknowns in this place, too many crosscurrents of emotion. The more time he had to think about it, the more certain he became that Hendrickson had an accomplice here. Of course, at Nichols's Twelvemile Roadhouse Hendrickson had a proven ally, for whatever reason. Even so, it was better to face known than unknown danger.

They were gathering their trail gear together, making ready for departure, when Wiseman had his first chance of the evening for a few words alone with Dreibelbis.

"Why did Edgar pull a shenanigan like that?" Wiseman asked.

"Well, Edgar is the clerk in the Circle office but, you know, the title 'deputy' goes along with the job. They've been shorthanded. I guess this was Edgar's one big chance."

"So he damned near gets himself and us killed in the bargain." Wiseman shook his head.

They were about ready to leave when Edgar walked up to Dreibelbis. "Anything I can do, just say it."

The nagging worry of moments before returned. "You can keep an eye on our back trail, Edgar. See that no one has a mind to follow us tonight."

With many a helping hand they were soon on their way. Fortunately it was not a long haul back to Nichols's Roadhouse, for everything was doubly trying now and slower. Hendrickson was made to run every step of the way ahead of the team, which meant frequent stops. The dogs, worn out by the rigors of the last three days, could manage no more than a weary trot.

Dreibelbis watched the team anxiously, fearful some of

the dogs might drop in their tracks. He was thankful for the full moon bathing the land in a light almost as bright as day, permitting him a clear view of each dog and of Hendrickson up ahead on the trail. Wiseman's pounding footfalls behind him were reassuring. At regular intervals Dreibelbis and Wiseman changed places on the runners. These changes kept them fresher than Hendrickson, an edge they would be sure to need.

Thus they inched through the night, the moon bright overhead, the frozen breath of dogs and men like a flowing streamer above and behind them. When at last the indistinct shape of the roadhouse loomed up ahead, neither men nor dogs had ever seen as welcome a sight.

NICHOLS'S ROADHOUSE

The dogs, normally Dreibelbis's first concern at the end of a day's travel, had to wait briefly this night. The roadhouse door had been thrown wide even as the team came to a halt. The deputies thrust Hendrickson ahead of them into its welcome warmth.

Nichols was ordered to place two chairs near the large drum heater in the combination dining and living room. Hendrickson, his trail gear removed, was shoved into one of the chairs. His arms were placed behind the back of the chair and his wrists manacled, following which leg irons were quickly adjusted. Only then did Wiseman remove his own trail clothes. He took the other chair, placed its back to one thick log wall, and settled down with rifle across his lap, facing Hendrickson ten feet opposite. Just as the last of these arrangements was completed, the door closed on Dreibelbis as he hurried out to care for the dogs.

"Him no let me unhitch dogs," said the Indian helper, pointing to Bingo who, despite his weariness, was hackled all the way to his tail.

"Never mind," answered Dreibelbis, glad to see that water was already simmering in a huge dog pot. Leaving

Photo by Frank D. Hengesbaugh

Handcuffs, which once confined the Blue Parka Man, and badge of Deputy Marshal George Dreibelbis, now the property of pioneer Frank P. Young

the preparation of the food to the helper, he unhitched the dogs, worried by the sight of their hobbled movements. He kneaded sore and stiffened muscles, starting first with Bingo, and did not finish until the food was ready.

Dreibelbis was concerned with many things as his fingers worked gently to relax the sore muscles. First was the dogs themselves. The fate of the deputies and their prisoner rested largely with them. If they lost any of the dogs on the return trip, their survival could be endangered. At best, the return would require an extra day, slowed as they would be by their prisoner.

Important as these considerations were, however, there were other matters of more immediate concern. Foremost among these was Nichols. How could he and Wiseman keep an eye on both Nichols and Hendrickson simultaneously and still permit themselves to get the rest they desperately needed? They had no charges that would stand against Nichols, so he could not be manacled like Hendrickson—the easy solution. The potential danger he represented, however, could not be discounted. It was unlikely that he would take any independent action. But if the opportunity should arise for him to help Hendrickson, with no danger or retribution attached to himself, he could be counted upon to act quickly and maliciously. Finally, there was the native helper. Where did his loyalties lie? Had he, like Nichols, forged any bonds with Hendrickson?

Dreibelbis waited until the food had cooled and the dogs were fed. The helper retrieved the quickly emptied pans, then turned to the marshal. "I go now," he announced.

"Tell my partner I'll be in soon," Dreibelbis replied.

After a final pat for each dog, he stepped out into the moon-bathed night. He walked on snow so hard-packed by the recent wind that his steps left no print. He climbed to the top of the knoll above the roadhouse overlooking the trail they had covered so recently. Not a sign of movement disturbed the vast solitude. Then he retraced his steps to the roadhouse.

"What kept you?" Wiseman asked. And almost in the same breath, "How is it with the dogs?"

Dreibelbis took Wiseman to one side. It was no good letting Hendrickson know any more of their problems than necessary.

"They're in bad shape. Probably need a day's rest. No telling until morning."

"I don't like it. This weather is damned changeable. The wind could blow up again overnight, and this place—it gives me the willies."

"We'd better make the best of it and get all the sleep we can. We'll need that, no matter what."

"Have you thought how we're going to do that?"
Wiseman demanded.

"I have an idea. You're safe enough there against the
wall. I don't know if I would be in the bunkroom, so I'll
just throw out my bedroll here where you can keep an eye
on me and Hendrickson at the same time."

"I could do with some sleep, too, you know."

Dreibelbis laughed. "We'll trade off. It won't be too
bad."

Nichols had spoken scarcely a word. He wandered back
and forth irresolutely between bunkroom, kitchen, and
dining room, casting occasional surreptitious glances at
Hendrickson. Now, as he observed Dreibelbis laying out
his bedroll on the floor, he could restrain himself no
longer.

"The bunkroom is cleaned up and waiting. You like
sleeping on the floor?"

"Your bunks are too soft, Nichols. Gives me a back-
ache."

That was wasted humor on the roadhouse keeper. He
stamped out of the room and was not seen again until the
following morning.

All of these arrangements and the low-voiced conversation between Dreibelbis and Wiseman had been observed
by Hendrickson. He detected in his captors a sense of
uneasiness beyond mere caution, and he diagnosed part
of it correctly. He had studied the team carefully at each
of their forced stops and knew how close the dogs were to
collapse. He guessed there would be no travel tomorrow.

"Sleep well," he said mockingly to Dreibelbis. Adjust-
ing himself as best he could in his uncomfortable position, he was soon asleep.

The deputies probably had less sleep than Hendrickson. Every two hours they traded places. The con-

stant vigilance while they were on guard, rather than making sleep easier when their turn came, seemed to defy sleep. Dreibelbis was relieved when the pretext of sleep was over and it was time for him to check on the dogs.

Although it was evident that two of the dogs could never stand the trip without another day's rest, none appeared permanently crippled. The night's rest, Dreibelbis thought, had done more for the dogs than for himself. He kneaded the still sore muscles of each dog in turn. When he was finished he felt certain that all of them would be ready next day. The knowledge was a relief, one less source of worry. The weather, too, seemed to be favoring them, for the temperature was steadily dropping. Not a breath of air stirred here on the valley floor or atop brooding Twelvemile Summit. Dreibelbis returned to the roadhouse to begin what on reflection was one of the longest days of his life.

Several times Hendrickson was released from his chair to relax cramped muscles. Twice, under the rigid vigilance of both deputies, he was permitted to take short walks between the roadhouse and the dog barn. On these occasions Dreibelbis sat with drawn rifle on the top steps of the roadhouse porch while Wiseman stood opposite, in front of the barn, where he could not only observe Hendrickson but also any movement within the roadhouse. Only at mealtime was Hendrickson permitted to have his hands manacled in front. The cuffs he wore could be a strangling weapon for one of his strength if slipped over a man's neck. The deputies could not afford such a risk.

Despite the walks with Hendrickson, mealtimes, and repeated visits to check on the progress of the dogs, the day passed with agonizing slowness. It was still only one o'clock that afternoon when Dreibelbis decided to postpone no longer the showdown between Hendrickson and Nichols. He had delayed this long out of an increasing sense of futility concerning the whole affair. He was not to be disappointed.

Nichols was permitted to clean up after the noontime meal. The helper returned to his seemingly endless job on the outdoor wood pile. Then Nichols was told to take a chair facing Hendrickson. He did so, studiously avoiding Hendrickson's challenging grin. Dreibelbis allowed Nichols to sweat for several seconds before speaking.

"You told us, Nichols, that you cooperated with Hendrickson because your life was threatened."

Nichols, although white-faced and trembling, was quick with his reply. "You said that, not me."

"Let's get this straight. You cooperated freely then with a man you knew to be the Blue Parka Man?"

"I told you, I didn't know who he was. You can believe it or not, but it's the truth." Nichols cast a supplicating glance at Hendrickson.

Hendrickson had been enjoying the whole proceedings—the obvious fear of Nichols and the frustration of the deputies. He laughed after the last exchange and spoke for the first time. "He's right, boys. Nichols didn't know me from Adam."

Right then Dreibelbis knew that he had lost the game. With Hendrickson's support, Nichols's answers became increasingly flippant. Not, Dreibelbis guessed, that it made much difference. Obviously Nichols had already decided that he had more to fear from Hendrickson than the worst the deputies could do to him.

Dreibelbis knew he would never learn what plans Hendrickson had in mind, whether it had been the robbery of the Birch Creek mines as Sid and the other owners feared. And what had Hendrickson promised Nichols? Whatever that was, it was probably only what Nichols had wanted to hear rather than any sharing of Hendrickson's real intentions. So, after all, what had he hoped to learn?

Dreibelbis turned from Nichols to look at his prisoner. Hendrickson must have been reading his mind, for he said, "What good would it do you to know?"

The question was never answered, for at that moment the helper burst through the door with a startling announcement. "Dog team come. Him come fast."

Dreibelbis did not bother with parka or gloves. Grabbing his rifle, he headed for the door, Nichols and the helper at his heels. Wiseman would know what to do with Hendrickson.

When Dreibelbis flung open the door, the team was less than fifty feet away, moving fast and with no apparent intention of stopping. Dreibelbis waved an arm. When there was no slackening of the team's speed, he jumped off the porch, rifle held aloft. Now the team was almost abreast. It was a long string of twelve dogs. One passenger rode the sled, so heavily bundled in fur robes as to be unrecognizable. The man on the runners, parka hood drawn close against the cold and the rush of wind, raised a hand in the barest of acknowledgments and the team was past.

Nichols's string of invectives did not end until he had to stop for breath.

Dreibelbis added a few choice sentiments of his own before asking, "Know who that was, Nichols?"

"Beats me. Never saw the cheapskates before."

Dreibelbis turned to the helper. "Ever see that team?"

"Maybe yes, maybe no." It was the stock answer he had expected from the native. He threw up his hands in exasperation.

Dreibelbis was spared an explanation when he reentered the roadhouse. Nichols had preceded him and was explaining for a second time what had happened.

"Strange way for a man to act on the trail," Wiseman said to Dreibelbis later that evening.

Dreibelbis agreed that it was, but Wiseman could see that his partner's thoughts were obviously elsewhere.

Dreibelbis tried to concentrate on all the logical explanations for the driver's actions. It was late in the afternoon and darkness was quickly approaching. No one in his right mind wanted to be caught on the summit in the dark. But most important, it was simply not the way of the country. A man was never too hurried to say hello.

10

SURVIVAL

From the outset the deputies had decided that the only way to handle Hendrickson on the long return trek to Fairbanks was to keep him so tired that, even if an opportunity for escape arose, he could not take advantage of it. They had no illusions that their two-to-one advantage and the irons binding Hendrickson were any guarantee of safety. Thus the morning's start found Hendrickson running ahead of the team with infrequent opportunities for rest.

Despite their predawn start, the sun had risen long before the summit was reached. But the men were hardly conscious of it. They had toiled up the tortuous ascent in the shadow cast by the mountain, and on reaching its crest they were nearly blinded by the dazzling glare of sun on snow. For as far as the eye could see in every direction there were limitless stretches of blazing whiteness, relieved only in distant valleys by stands of spruce.

Their rest atop the summit was brief. As they gazed on the view below them, they were less impressed by its beauty than the surprising change that had taken place. The storm that had raked Twelvemile and Eagle summits had dumped a heavy fall of fresh, powdery snow on the southern mountain slopes. Even from this distance it was clearly discernible, and it would mean slow, painful traveling when the valleys were reached. Quickly, rough locks were tied to the sled runners. They were ready to begin the risky descent.

It was two hours later, in a valley protected from the main force of the wind, that the newly fallen snow became a problem. The dogs floundered belly deep, lunging, no longer able to pull in a sustained effort. Dreibelbis called a halt while snowshoes were broken out. Hendrickson, forced to break trail, took the brunt of the work. Wiseman followed some twenty feet behind, then the hard-straining team with Dreibelbis pushing as best he could but sinking knee-deep with every staggering step.

They crossed a narrow ridge, temporarily spared the leg-cramping ordeal of trail breaking, descended into a wider valley, and there found for the first time the freshly broken trail of another team. It could only be the team of the day before, but it had taken a more precipitous route off the mountain.

Dreibelbis knelt beside the trail, studying the declivities left by sled runners, the footprints of dogs and man, and the consistency of the surrounding snow. He rose at last. "Eight hours old, I'd say. They probably camped behind that rise. We're about a day's travel behind them."

"Well, I'm glad we're behind and it's them breaking trail," Wiseman said with an effort at cheerfulness.

Dreibelbis had to agree with that, but he wondered when they would catch the team ahead. It was sure to be at a time and place of the stranger's choosing. Better that the team was ahead rather than lurking at their heels.

Dusk was beginning to fall when they saw the deserted cabin where they planned to spend the first night. The

tired team came to an automatic halt at the cabin. The
men stopped motionless in their tracks for several moments, on legs numbed to all feeling. At last Wiseman
knelt down to release his snowshoe bindings, then did
the same for Hendrickson. Dreibelbis and Wiseman
moved through the evening chores in a kind of slow-
motion pantomime. Hendrickson had asked for a robe,
and he huddled down in a corner, unmoving until dinner
was ready.

Sleep was the predominant thought for all three men,
and it was the most carefully arranged detail of the night.
Robes were laid out on the floor for a single bed to accommodate both deputies and their prisoner. Hendrickson, handcuffed and leg-ironed to Dreibelbis, slept
in the middle. Wiseman was on the other side nearest the
corner of the cabin where the guns were stacked. He was
the only one of the three with any mobility and could
reach the guns quickly if necessary, but that was hardly
likely. Dreibelbis and Hendrickson were powerless to
move or to change position. Had they not been so completely exhausted by the exertion of the trail, probably
neither could have slept. But they did sleep, and morning, when it came, was much too soon.

The trail the next day became progressively worse, the
snow deeper the farther they traveled. Had it not been for
the team ahead, they might not have made it. The trail
they followed had firmed somewhat overnight, so there
was a bottom of sorts to it. Not that it made the snowshoeing ahead of the team easier, for the slight crust that had
formed had to be broken with each plodding step, but the
team found firmer footing.

Dreibelbis imagined what torturous travel it must be for
the other team. The signs were plain to read. The distance
between stops became shorter. At one point the dogs had
laid down. Flecks of blood on the snow showed how cal-

lously the driver had used his whip. Finally he had taken out snowshoes to break a trail, and the team had followed.

All this Dreibelbis read easily, but he had no way of knowing the man's intentions. The valley they were in narrowed perilously in several places, affording every opportunity for ambush. Nothing had happened. Each time they emerged from the defile to find the trail they were following, stretching ahead endlessly. Tired, his team failing, the man had lost whatever advantage he once possessed. Why had he waited this long to strike, if indeed that had ever been his intention?

And what of the second man Dreibelbis was so sure he had seen on the sled? Was that also a phantom of his imagination? Nichols had seen him, too. There had to be a second man, but in all this time they had seen but a single set of tracks. If there was a second man he had never left the sled.

They had just resumed their plodding progress after the noon stop when they found the dead dog alongside the trail. It had been cut out of its traces, not even spared the decency of a bullet, apparently bashed in the head by a rifle butt. Whip marks were plainly visible. Dreibelbis, angry at the needless brutality, bent over to feel the dead animal. It had already frozen stiff in the bitter temperature.

"Four hours at most," Dreibelbis said, in response to Wiseman's question. "We'll catch him before the night stop."

Dreibelbis turned on Hendrickson. "You know who this is?" he demanded. For once there was no mockery in Hendrickson's glance, and his face was gray with exhaustion. He shook his head wordlessly, and Dreibelbis was convinced. At last Dreibelbis had to admit to himself that he had probably been wrong. But if they were not trailing Hendrickson's accomplice, then who?

Two hours later they found out. As they topped a slight rise in the trail, they saw the exhausted team they had followed for so long. A man sat dejectedly on the bow of the sled facing ahead toward the unbroken expanse of snow. He was not yet aware of the approaching team. No sign of a second person was visible.

"Stand there ahead of the team," Dreibelbis told Hendrickson, "and don't make a move."

Dreibelbis and Wiseman, rifles drawn, staggered up the trail toward the cowering team that, bellied down in the snow, had refused to move farther. Something alerted the man on the sled. He turned around and stumbled to his feet. Seeing two armed men bearing down on him, he threw his hands in the air.

"Don't shoot," he pleaded, stepping away, then falling backward in the knee-deep snow. The next instant Dreibelbis towered above him, rifle barely two feet from the man's face.

"Where is your partner?" Dreibelbis demanded.

"What partner? God, man, I got no partner."

Dreibelbis's rifle barrel pressed closer, and the stranger squirmed deeper into the snow trying to avoid it.

"Hold it," Wiseman said. "He might just be right." Wiseman finished poking through the sled load and whistled in amazement. It contained a small fortune in furs. "No room here for a passenger."

It was hard for Dreibelbis to acknowledge that he had been wrong on all counts but, looking at the high pile of furs on the sled, he knew that he had been. His brief glimpse of the sled at the roadhouse had led to a second mistake. Dreibelbis swung back to the stranger.

"Just who the hell are you, and why your hellbent rush that you kill off your team to get there?"

"Name's Jacobsen. Fur buyer. If I don't get those furs to market fast, the price can drop and wipe me out."

"I hope it does. And here's some advice you can believe. If I hear of you laying whip on a dog again, I'll hunt you down and wear out the whip on you."

Jacobsen was ordered to make camp where he was, rest his dogs, and proceed next morning to the cabin over the next ridge where the deputies would spend the night. He could stay there until the help they sent reached him. Dreibelbis figured it was more consideration than the man deserved.

That night was spent as the last one had been, Hendrickson and Dreibelbis chained together in a shared misery that only the day's suffering made endurable. Fatigued, cramped muscles cried out in constant pain that the night's rest could not relieve. There was still another day to go—and on an unbroken trail now that the other team had been left behind. The dogs were as close to physical collapse as the men. No longer did Hendrickson pose the major threat. It was the trail now, and it promised as slight a chance for Hendrickson as for his captors.

None of them would have made it, they were quick to admit in later days, except for the vagaries of the recent storm. Soon after their morning start it was apparent that the snow depth was less, and it continued to lessen throughout the morning. Finally they were able to dispense with the snowshoes. They stopped. Hendrickson, still in the lead, fell to his knees. Dreibelbis dropped beside the sled, back propped against it for support. Wiseman sat in the middle of the trail, too exhausted to move the few feet to the sled. The dogs dropped to their bellies or lay on their sides, breathing in shuddering gasps. For ten minutes they remained nearly motionless in grateful relief despite the forty-below-zero temperature.

When they started again, their progress was painfully slow despite the flat, hard trail. It was as though they had

left the last of their strength in the snow-clogged trail behind them. It was early afternoon before they finally reached the abandoned Cassier Roadhouse, with Fairbanks only fifty miles away. And it was there that they met Tom Gibson, returning afoot to his isolated camp.

He stopped and waited as the group drew nearer. Then he saw with a shock of surprise that the man in the lead was Hendrickson.

"Hey," Tom shouted, "what's up?"

Hendrickson gave Tom a nod of recognition but did not pause. He was breathing hard and half-turned, thrusting out his wrists to show Tom the irons.

Wiseman shouted, "Keep away from that man," and at the same time pulled a rifle from its boot atop the sled load.

Dreibelbis, who had dropped behind the sled, struggled up alongside. "It's all right," he said to Wiseman on recognizing Tom. "I know this man. Keep going, Frank. I'll catch up."

The two friends' conversation was brief. Hurriedly Dreibelbis recounted the events on Birch Creek. "Stop by when you're back in town and I'll fill in the details."

"I'll do that," Tom promised. He watched until Dreibelbis disappeared down the trail. No one knew better than he what qualities it had taken to make this trek. And a last element was luck. On a trip such as this with all its possibilities for disaster, add a big measure of luck.

It was ten miles farther along the trail, with their goal so tantalizingly close, that the first dog dropped. It lay in the snow, its tail barely moving. Gently Dreibelbis unhitched the dog and placed it on the load, arranging it as comfortably as he could. They traveled another ten miles, much slower now with only five dogs to pull and the extra load. The deputies had thought of but one thing in the last hours. To reach Cleary City in time to catch the evening stage for Fairbanks. It was the lodestone that kept them struggling, forcing aching muscles beyond endurance. Then the second dog dropped in its traces. The team was finished. Dreibelbis had already asked more of the dogs than flesh could give. Slowly he lifted the second dog onto the sled, throwing a robe over both animals. He unhitched the other dogs so they could find a measure of comfort.

Hendrickson and Wiseman watched Dreibelbis, grateful for the momentary relief but knowing that it could not last.

"The dogs and me will make out here till morning. You'd better get moving. With luck you can still catch the stage."

Stumbling, falling, in pitch darkness the final miles, too tired even to curse, Hendrickson and Wiseman kept going, and luck was with them. At Twelve Below Roadhouse, even before Cleary City was reached, and before they were close enough to see the welcome roadhouse lights, they heard the blasphemous voice of Al Roberts, the stage driver, swearing at his horses. They could not believe it. The stage never traveled beyond Cleary. But on this particular night Al, with a full load of passengers bound for the roadhouse, had made an exception.

They made it with nothing to spare when Al finally heard Wiseman's last despairing shout. Thankfully there were no other passengers. Wiseman helped Hendrickson mount the steps into the stage. He never replaced the leg irons; it was an indignity he could not have imposed that night. The trail, and especially such a trail as they had traveled together, created a kinship that had to be respected. Hendrickson had more than earned that respect.

The stop at Cleary was brief. Al was told to call Marshal Perry with the news of Hendrickson's arrest and to say they would be arriving on the Cleary stage. There were no passengers at Cleary, not that it would have mattered: Wiseman would have ordered them left behind. Al brought fresh foot warmers and, despite a curiosity he could not hide, asked no questions. Their speed on the trip to Fairbanks that night set a stage record that was never equaled.

Al wasted no time with a stop at his regular station. Instead he pulled up with a flourish directly in front of the jail. The wheels of the stage barely stopped when the jail door burst open and two armed deputies rushed to open the stage door.

Every light in the jail was burning. The sudden brilliance blinded both men as they emerged from the darkness of the stage. Quickly they were shoved inside the stifling jail interior. Not only was Marshal Perry there in person, but so was Reynoldson, the Chief Deputy Marshal, and every deputy, jailer, and guard on the staff, plus the entire newspaper contingent of the town—publishers, editors, and reporters. The small jail bulged with the noisy, curious crowd. Across the room, trying vainly to reach him, Wiseman saw Charlie Dreibelbis, George's brother, with a worried expression on his face. Wiseman gave him a reassuring wave and saw Charlie relax. Then the reporters were at him. He answered questions for an interminable half hour before refusing to be bothered further.

Wiseman had not seen Hendrickson since they entered the jail. His prisoner had been whisked away to the cell blocks. Now he headed in that direction.

Hendrickson, heavily ironed, lay on his bunk facing the wall. He had refused to answer questions from reporters. Wiseman admired that, wished he could have done the same. He came to the cell bars and spoke Hendrickson's name. Recognizing the voice, Hendrickson turned around to face him.

Wiseman nodded. "That was a tough trail," he said. "You were a good prisoner." It was high praise from a man who rarely gave any. "I'll see you," he said as he turned to leave, never dreaming under what circumstances their next meeting was to occur.

Hendrickson's recapture was a seven-day wonder. It invoked all the excitement of his summer's forays. Again the Blue Parka Man was the focus of endless conversation and conjecture. Would the authorities actually be any more successful in restraining him this time than they had been before? Many doubted it.

11

JAILBREAK

The year 1906 started as 1905 had ended. January provided no relief from the unrelenting cold of the preceding three months. When on rare occasions the temperature rose to twenty below, it seemed like shirt-sleeve weather. Heavy parkas were left in closets, and hardier souls could be seen bareheaded about town, but these periods of relief from the intense cold were as brief that winter as they were few.

The deep cold creates a world of its own. As the temperature drops, the air becomes unbelievably still. A breathless hush hangs over the land. A person abroad in this weather can easily envision himself as the only inhabitant of a desolated earth.

At times it might seem to be snowing, but it is too cold to snow. The fine particles drifting from the sky are frozen shreds of moisture crystalizing out of the air, coarse and rough as sand. With no breath of air stirring for weeks on end, frost builds up like garlands as much as two inches high on branches, fences, and telephone lines. Limbs and wires sag under the heavy accumulation and often break. Even the sturdy spruces become so brittle in the prolonged cold that their branches can snap at a touch like dry twigs.

The most ordinary aspects of living become distorted in peculiar ways. The ice fog hanging like a dense pall over the town hides all traces of the sun that shines unseen. Open a door and the fog rolls in like a cloud.

There was no diminishment of activity the winter of 1906, although it was definitely channeled into action with an indoor setting. Indeed there was a desperate urgency for social life to avoid the winter's prolonged confinement. Socials, dances, and card parties by clubs, churches, and lodges drew overflow crowds, to say nothing of the gaudier entertainment provided by the saloons and dance halls. In addition the prize-fights at Century Hall always attracted a capacity audience.

For the jail inmates there were none of these diversions for their even drearier days. The cells were always cold, clothing and bedding were inadequate, and the lights that shown so brilliantly the night of Hendrickson's return were reduced to a few dimly glowing bulbs that left the interior in perpetual gloom.

Charles Hendrickson's visitors were few that winter. Leroy Tozier, his lawyer during his earlier trial, was an infrequent visitor, completely lacking in his former joviality. He was probably much more interested in obtaining the fee for his previous services, Dreibelbis surmised, than in representing his client at a second trial.

Hendrickson was not the only well-guarded jail inmate. There was another prisoner by the name of Thomas Thornton with a record of one escape from the federal jail. The two men were as dissimilar as possible, but in late January an event occurred that would link their futures inseparably.

Thornton had been released from the Fairbanks jail on June 1, 1905, after serving a one-year sentence for grand larceny following the break-in of a Fourth Avenue cabin. One month after his release, he was charged with horse stealing, the only known crime of its kind ever perpetrated in Alaska. In describing this bizarre event, the *Fairbanks Times* reported, "On July 2 in company with a man named Hansen, he led two halters away from Cleary City at the end of which were two fine horses."

It was not until August that Thornton was arrested at Fort Gibbon with the assistance of the Army garrison. He was lodged in the Fairbanks jail, from which he escaped barely a month later. This time he was apprehended at Eagle, almost at the Canadian border. On November 30 he was back again in the Fairbanks jail which, if it lacked most of the amenities of home, must by now have become almost as familiar.

In contrast to Hendrickson, Thornton was regarded by those who knew him as a shrewd but not too bright individual, sullen and with a vicious temper. Although an Englishman by birth, his complexion was dark, with a beard so black that he looked to be always in need of a shave. He was thirty-eight years old and of medium build.

Checking on the prisoners was only one reason for Dreibelbis's visits to the jail. He was also keeping an eye on the jailers. Neither of these chores was his responsibility or rightly his concern. But when he was not on the trail and time permitted, he felt an obligation to do so. Of late he was particularly concerned about one of the two jailers on the day shift, Jim Gallagher. It had been necessary to reprimand him several times for indifferent treatment of the prisoners. As senior of the two guards, it was his responsibility to see that the food was of proper quality, that it was served hot, that the cells were kept reasonably clean, and that blankets were regularly aired. None of these duties, Dreibelbis observed, was being adequately performed.

By contrast the second guard, Pete Peterson, was a conscientious, competent employee. He did what he could to make the prisoners' lives more bearable, while Gallagher taunted them and treated them with abusive contempt. The day Dreibelbis overheard Gallagher telling the prisoners that if only he could get at them he would single-handedly give them the beating they deserved, he went directly to Marshal Perry and asked that Gallagher be replaced. It was wasted breath. Perry was more amused than concerned. He was of the opinion that prisoners were not to be coddled, and he was sure that Gallagher was doing a commendable job. In anger and disgust Dreibelbis stormed out into the late January gloom that passed for daylight, vowing never again to set foot in the jail unless ordered to do so.

It had turned steadily colder in the early morning hours of January 29. The night guards at the jail, happy that their long shift had ended, were donning heavy parkas and mufflers while exchanging casual gossip with Peterson and Gallagher, just arrived for their tours of duty.

"Those are the lucky ones." Gallagher pointed at the prisoners in derision. "Look at 'em. Free food and lodging. They don't have to go out in that damned cold."

The other guards made no answering comments. Gallagher's constant riding of the prisoners was becoming tiresome. Besides, there was no way a man in the light clothing permitted could stay comfortable in the partially heated cells.

The night guards pulled fur hats well down over their ears and waved before opening the outer door.

Half an hour later, Peterson was serving breakfast to the prisoners when Thornton threw a handful of pepper in his eyes. Although blinded and taken completely by surprise, Peterson grabbed Thornton and hung on desperately. Thornton had unexplainably acquired a case knife, and in the brief fight that followed, Peterson was stabbed twice in the region of the heart and again in the back. One slash severed an artery.

As soon as Thornton threw the pepper, Hendrickson, with the litheness of a big cat, sprang through the cell door at Gallagher. With two swift blows he knocked him down and out. He had managed to slip his leg shackles unobserved during the morning. Now he quickly pulled on one of the guard's parkas and shoved his feet into a pair of heavy boots. He threw one hurried glance at the two men still struggling inside the cell, in time to see Thornton deliver his second slashing stroke at Peterson. He hesitated only a second longer when his glance fell on the locked rifle rack. Gallagher was beginning to stir, and it would mean ransacking the guards' pockets to find the key to the rack. Discarding the idea, he threw wide the jail door and disappeared into the darkness. Hendrickson's only concern was to reach his prearranged rendezvous as quickly as possible. Everything he needed awaited him there: rifle, warm trail clothes, food, and trail gear.

The jailbreak had been thoroughly planned to the last detail with the active help of friends on the outside. There was no other explanation for the events of that morning—for the help Hendrickson needed to free himself from his leg irons, for Thornton's possession of the case knife, and for his subsequent, successful disappearance. Most convincing proof of all could be found in the weather. No one in his right mind would engineer an escape in this temperature without being thoroughly prepared for it.

Hendrickson had already vanished into the pitch blackness when Peterson fell helplessly to the jail floor. Thornton, still in his leg irons, stepped awkwardly across him. It never became clear why he had been unable to shed his irons as Hendrickson had done. Quickly he wiped the blood from the knife on Peterson's jacket, shoved it into an arm sheath, and hobbled across the room. He grabbed a parka from a chair and struggled into it, then headed out the still open door. Only the other jail inmates witnessed the escape. The early hour, the darkness, the way people were unrecognizably bundled against the cold all aided the prisoners' flight.

None of the other prisoners made a move to escape, nor did they give any assistance to either of the jailers. Unmoved witnesses to the swift attack, they huddled back in their cell, prisoners not of bars but of the cold.

Gallagher regained consciousness in a matter of minutes. He was groggy but uninjured. He took one look at Peterson and the widening pool of blood seeping out across the jail floor and rushed out into the cold, heading for the marshal's office.

George Dreibelbis had arrived early at the office that morning, as was his invariable custom. He was followed shortly by his brother, Charlie, who had barely removed his parka when a wild-eyed Gallagher, coatless and hatless, burst through the door. In the short distance between the jail and the marshal's office his cheeks and ears already showed the marks of frostbite.

When Dreibelbis had calmed down the hysterical Gallagher enough to learn what had happened, the other deputies began arriving. One he dispatched immediately to the jail to take charge there and to see what could be done for Peterson. He sent a second deputy across the street to the nearest doctor with orders to get the doc to the jail in his long johns if necessary. A third deputy was busy on the phone raising the other members of the staff.

Dog-team posses were quickly organized, one of them led by Charlie Dreibelbis. They fanned out from town on every trail. By the time Marshal Perry arrived at the jail, all the wheels had been set in motion. Nothing was left

Charles Bunnell Collection, Archives, University of Alaska,
Fairbanks

Along the Valdez Trail

for Perry to do but curse and fume and, later that morning after the news spread, try unsuccessfully to mollify the again virulent press.

Something went unaccountably awry with Hendrickson's plans. Whoever was to have kept the rendezvous with him that morning failed him. Where that meeting was to have taken place and with whom was never learned. It can only be surmised from Hendrickson's subsequent movements that the meeting was to have been somewhere on the southern outskirts of town, and that when he found no accomplice and no trail outfit he felt it was unsafe to remain. He struck out afoot along the Val-

Water wagon with stove, used for house-to-house deliveries in Fairbanks

dez Trail wearing only the clothes that had been on his back at the time of his escape. It was an unaccountable act of desperation, totally inconsistent with his earlier and later actions, as though he sought certain disaster. For twelve tortuous miles he kept going, shaking with cold, at times only half-conscious. At length he found a deserted cabin and staggered in, undoubtedly hoping to find a stove and warmth. But it was empty. It was there that Charlie Dreibelbis found him not long after, huddled in a corner of the cabin, shivering uncontrollably.

Even before Charlie Dreibelbis had taken the trail, an ugly crowd had begun forming before the jail. There had been muttered talk of lynching. Now, as Charlie neared the jail with his prisoner, he figured the mob would be drunk enough not to care which prisoner they strung up.

Charlie had his team going full tilt when they rounded the corner half a block from the jail. With a shower of ice kicked up by the five-pronged brake, he brought the team to a stop just at the jail entrance. He was holding the back of the sled with one hand and waving his rifle with the other. "This is Hendrickson, boys," he shouted. "Keep back. Stay clear."

Guards whisked Hendrickson from the sled and through the jail doors before there was any reaction from the lynch mob. Most of its members had retired to nearby saloons, leaving a few frequently rotated men to keep watch on the jail. Now, alerted to Charlie's arrival with the prisoner, angry men poured into the street, storming toward the jail. The deputy on guard at the jail entrance threw up his Winchester at the first move from the fast-forming lynchers, joined by Charlie Dreibelbis as soon as he had seen his prisoner safely through the jail door. For tense moments it was an explosive situation. Charlie always claimed that it was only the cold that prevented wholesale tragedy. There is nothing like the chill of fifty

below to quell passion and to cool tempers. It was fortunate also that the prisoner was Hendrickson, not
Thornton. Gradually the crowd dispersed to the more
convivial atmosphere of the saloons.

Tempers in the town continued to seethe. Not only was
it fortunate that the prisoner captured the day of the jail
break was Hendrickson, it was doubly fortunate that a
lengthy time elapsed before Thornton was apprehended.
Two days after the attack on Peterson, the *Daily Times*
reported: "The prisoner Thornton, who escaped from the
Fairbanks jail several days ago, is still at large. It is believed he is being harbored by friends and is armed. The
officers are hot upon the scent and expect to have him
before morning." The story continued with a report on
Peterson's condition: "Peterson was operated on today
and there is one chance in five hundred of his recovery."

Despite the confident opinion expressed by the *Times*
concerning Thornton's capture—and ardently shared by
Marshal Perry—Thornton was not captured until almost a
month later, on February 26. It was under such unexpected circumstances that it quelled all further thought of
mob violence.

The *Times* was correct in one assumption: Thornton
did find safety with friends. His leg irons were removed,
he was assisted out of town, and he was provided refuge
in a woodcutter's tent some fifteen miles below the town
of Chena and a mile back in the woods on a seldom-used
trail. It was there, acting on a tip from an undisclosed
source, that Charlie Dreibelbis and a fellow deputy found
him in the early morning of February 26.

Warned that Thornton was well armed, the deputies
advanced carefully on the tent. They took well-protected
positions before Charlie hailed Thornton and ordered
him to come out with his hands in the air. There was no
answer. Charlie shouted three more warnings, finally that

they were opening fire if Thornton did not show himself. Still not a sound or sign of movement from the tent. Charlie fired a shot through the top of the tent. No response.

"Something damned funny here," Charlie said. "Keep me covered while I try to get inside."

Charlie approached not from the front but from one side of the tent, a drawn hunting knife in one hand. The second deputy continued firing at intervals. With a leap and a slashing motion Charlie ripped open the side of the tent and dove inside. The next instant he was shouting at the top of his lungs for the other deputy.

Thornton was stretched out on the floor of the tent, bleeding profusely from four savage self-inflicted gashes in his throat that extended from the lobe of his left ear to under the chin. A dull, bloodied pocketknife lay where it had fallen from his right hand.

Charlie bandaged the ugly wounds as best he could to stop the flow of blood. The deputies carried Thornton to the dog sled and rushed him to the Fairbanks hospital. Thornton was so weak from loss of blood that Dr. Cassells, who attended him, said his survival was a miracle. Thornton remained in the hospital for three weeks before he could be returned to jail.

Spring finally did return to the North that year of 1906. The sun worked its miracle again. Beginning with only fractions of a minute, daylight was soon gaining at the rate of seven minutes a day. The pale green of the pussy willow buds was the first tinge of color to promise spring, followed soon by the blossoming gray flowers, soft as velvet. The pussy willows were a false promise, for it snowed again while they were still in bloom, and the March winds were still a reminder of winter.

The ice finally left the Tanana and Yukon rivers, however, and crews were soon busy on the creeks around the

clock, shoveling the rich gravel from the winter dumps
into the sluice boxes. With the Blue Parka Man safely
behind bars, the cleanups began flowing in an uninter-
rupted golden swell into Fairbanks.

Indeed, the marshals were doing everything within
their power to insure that two escapes apiece for Hendrickson and Thornton was their absolute limit. As an
added precaution against further attempts, the leg irons
on both Hendrickson and Thornton were riveted in place.
No more chances were being taken with locks that had
proven so ineffective in the past. Moreover, the prisoners
were always under the muzzles of armed guards.

Pete Peterson's life hung in a delicate balance throughout that spring, but he eventually recovered. Perhaps it
should be said that he lived, for he was to walk stoop-
shouldered and with a decided limp from the day he left
the hospital. As for Thornton, he fared little better. There
was no hiding the ugly scars from his self-inflicted
wounds, and his right hand was useless when raised
above his head. For them perhaps spring never did arrive.

There was nothing about the dawn of May 22, 1906, to
mark it as a memorable date. Nearly eleven months had
elapsed since the devastating flood that had almost swept
Fairbanks out of existence. The town had recovered from
that catastrophe to grow even bigger and more vigorous.
It was the young giant of the North.

Spring was fast giving way to summer on May 22. It
was a still, hot day without a hint of breeze. Men were in
shirt sleeves; women carried parasols as protection
against the scorching rays of the sun, when such a short
time before they had been unrecognizably bundled
against the cold. People were not yet complaining about
the heat. Winter was still so recent that it was a pleasure

Charles Bunnell Collection, Archives, University of Alaska,
Fairbanks

The start of the great fire in Fairbanks, May 22, 1906

to swelter again. At three forty that afternoon business was brisk in the downtown section of Fairbanks, with no hint of impending tragedy. Twenty minutes later the town was a raging inferno.

The fire burned furiously until seven o'clock that evening, by which time four square blocks comprising the commercial heart of Fairbanks had gone up in flames. Fairbanks, however, was destined to survive. Flood had done its worst to destroy the town. Now fire took its turn, and it too failed. The embers had not cooled before there was talk of rebuilding.

The fire losses were not confined to the business district alone. The courthouse and the jail were also lost. Few people except for the marshal and his deputies were paying attention to the jail inmates as the fire raged, but as flames licked closer and closer to the tinder-dry log jail, the prisoners, leg-ironed together in single file, were led from the jail under the watchful eyes and triggered rifles of the deputies and guards. As hot brands from the approaching fire fell among them and the smoke-filled air caused their eyes to smart and water, the prisoners were led down Third Avenue to a large vacant lot on Wickersham Street. There they remained, confined to the very center of the lot, under the ceaseless vigilance of rotated deputies and guards, until four days after the fire when a new, temporary jail was completed. There was never less chance for a break. No cell or irons were as good deterrents to escape as determined men with rifles. Better opportunities were certain to come later.

Marshal Perry had also learned a few lessons from the harsh hand of experience. George Dreibelbis had been placed in sole charge of the jail inmates. As the town threw itself into the massive task of rebuilding, Dreibelbis was given the added responsibility of designing and supervising the construction of an escape-proof, permanent jail.

12

THE DEPARTURE

Fairbanks, after December 1, 1904, was headquarters of the court for the sprawling Third Division, covering the immense northern half of Alaska. Yet court sessions in Fairbanks were intermittent. Judge Wickersham administered a traveling court, his vast jurisdiction extending from Rampart on the Yukon to Dillingham and Valdez. Under these circumstances justice was not always swift or even certain.

The slow wheels of the law were further impeded the spring of 1906 when Judge Wickersham was called to Washington and detained there until late summer. By the time of his return, six months' business had accumulated before the court. Hendrickson and Thornton languished in jail until mid-August, following their January escape attempt, before they next appeared in court. It is surprising that no one seemed to believe they could be restrained from escape for that length of time.

The marshal's office was taking every precaution to preclude further escapes. The most effective single deterrent undoubtedly was the riveted leg irons that Hendrickson and Thornton were compelled to wear day and night. There was no manipulating cold steel. Beyond this, the permanent jail now under construction was designed to supposedly foolproof specifications. Although not completed in time to house Hendrickson and Thornton, it deserves special note because of the part it played in their plans.

Four cells enclosed in a steel cage were provided for the more difficult prisoners. The cell doors locked automatically, and when closed there was no escape except by sawing through Bessemer steel bars. Since the prisoners would be in full view of guards twenty-four hours each day, this was a highly unlikely possibility.

These elaborate detention plans were no secret to Hendrickson and Thornton. As events of the summer were to prove, they were probably as familiar with all details of the jail's design as the marshals themselves.

In Judge Wickersham's absence the Grand Jury was convened. It indicted Hendrickson for the January 29 jailbreak attack on James Gallagher, and Thornton for the brutal knifing of Pete Peterson. These indictments were in addition to the charges already pending against the prisoners.

There was still intense public resentment against Thornton. Pete Peterson's painful shuffling about town after he was finally released from the hospital was a constant reminder of the vicious attack. In Hendrickson's case, however, the public was quick to condone his part in the event. After all, he had attacked the bully Gallagher—whose reputation was no secret—and, despite the provocations, had done the guard no physical harm beyond what he obviously deserved. Hendrickson continued to have numerous admirers. His romantic reputation of the previous summer persisted, and many were outraged that a man should be held in jail for such an unreasonable length of time.

In early August Judge Wickersham returned to Fairbanks and a new session of court was convened. On the
afternoon of August 17, Hendrickson and Thornton were
brought into court to plead to the indictments returned by
the Grand Jury.

For their court appearance the riveted leg irons were
removed from the prisoners and "Oregon boots" were
substituted so they might walk to the courthouse. Oregon
boots were an ingenious contrivance considered to be as
escape-proof as any device yet designed. They consisted
of circular iron disks weighing thirty pounds each,
clamped around the ankles by means of an inside hinge
and secured by a bolt sunk in a small socket. This was
locked by screwing it tightly with a key inserted in the
socket. The key had a circular opening that was pushed
down over the bolt. Iron bars with the ends bent upward
were screwed to the soles of a prisoner's shoes. When the
shoes were on the feet, the upturned ends of these bars
were attached to the Oregon boots in such a way as to
hold them up off the ankles so a prisoner could walk for
short distances. At night a prisoner's shoes were removed
and the boots dropped down on the ankles, making it
completely impossible to walk.

Thus confined and under the strictest surveillance, the
prisoners made their long-delayed appearance before
Judge Wickersham. It had been a difficult summer of rebuilding, and the town was ready for any event that
promised excitement. Milling throngs crowded and
pushed for admittance to the overflowing courtroom. The
citizens of the town were getting their first glimpses of
the prisoners in many months and, for those unfortunate
enough to miss the occasion, the newspapers provided
minute details of the prisoners' appearance and every incident of the suspense-filled afternoon.

Both prisoners looked pale from their long confine-

ment. Hendrickson was described as having ". . . aged
wonderfully in the last few years and is stooped with
either worry or time setting heavy on his shoulders." Despite this appraisal and the Oregon boots hampering his
easy, flowing movements, he presented a strikingly hand-
some appearance. He wore a gray coat and vest and, although topping blue overalls, his clothes managed to
achieve a fashionable effect not lost on the women in
attendance. His sandy beard was neatly trimmed to a
point.

As the deputy clerk of the court read the indictment
presented against him, Hendrickson listened attentively.
He offered no comment. When Judge Wickersham asked
how he pleaded, he answered without hesitation, "Not
guilty." The mocking smile so characteristic in the past, if
it came less frequently on this occasion, was still present,
the same self-assurance. It was not lost on Judge Wickersham when he asked Hendrickson to enter his plea, nor
on George Dreibelbis watching the reenactment of a painfully familiar scene.

Thornton, obviously nervous and worried, presented a
complete contrast. His appearance was disheveled and
careless. He had grown a heavy black beard that almost
but not quite succeeded in hiding his scarred features. He
pleaded not guilty in an almost inaudible voice.

There was a sudden hush throughout the courtroom
when the second plea was entered, as though the packed
throng expected some spectacular development. That
was to await another day. The judge quickly remanded
both men for trial. The proceedings were completed with
disappointing speed, and the courtroom spectators
headed grumbling toward the doors they had entered
with such lively anticipation.

If there was any unexpected development, it had to be
that the prisoners were represented in court by Leroy To-

zier, the same lawyer who had acted for Hendrickson at his earlier trial. After Tozier's disillusionment following Hendrickson's first jailbreak, what could possibly have persuaded him to take the present case? Everyone agreed that it had to be money. If that was the answer, where did it come from?

Whatever his reward, Leroy Tozier was busy earning it in the two weeks that followed. In a round of meetings with the prisoners, District Attorney Harlan, and the court, Hendrickson and Thornton were persuaded to change their pleas to guilty in exchange for lenient sentences. The negotiations were carried out in strictest secrecy, so when it was unexpectedly announced that Hendrickson and Thornton would appear in court on Saturday, September 1, to amend their pleas, the newspapers and the general public were equally astounded. These events contributed to another overflow crowd when the courtroom doors opened, and now the audience was treated to all the dramatics they had expected but were denied at the earlier trial.

Almost as soon as the court convened, Thomas Thornton took the stand. Ash-white and trembling, he delivered a half-hour plea in his defense. A pin could be heard to drop in the tense silence as spectators bent forward to catch every word of his low-voiced account. Begging the court's indulgence, he said that he felt he must explain why he made his jailbreak. Jailer Gallagher's brutality and the intolerable jail conditions, he said, had driven him to it.

At this point District Attorney Harlan interrupted Thornton's account to say that he was taking advantage of the court's indulgence and to strongly deny the charges against the marshal's office. Thornton, to the delight of his audience, was permitted to continue. He voiced the highest praise for Pete Peterson, who, he said, had fought him like a man, and the deepest regret for the harm done him.

"I am not a vicious man," Thornton told the court. "Since coming North it seems that everything has gone wrong." Thornton's words were barely audible to the courtroom spectators, and it appeared to many that he was about to break into tears. "Judge Wickersham, I have a wife and family—and I would—like to see them—again." After these last words, he threw himself on the mercy of the court.

For a full ten seconds following Thornton's disclosures there was dead silence in the courtroom, followed by a low hum of comments. Judge Wickersham banged his gavel once before delivering his opinion. Possible mitigating circumstances and the prisoner's admission of guilt, he said, in no way diminished the seriousness of the offense. Thornton was sentenced to serve fifteen years in the federal penitentiary at McNeils Island, Washington.

Almost immediately Hendrickson's name was called. He rose and merely entered a plea of guilty. If Judge Wickersham had expected any remorse from Hendrickson, he was disappointed. In a sharper tone than he had used throughout the hearing, he pronounced an identical fifteen-year sentence for Hendrickson. This produced a loud, incensed murmur from the spectators. No one in the courtroom that day or later believed Hendrickson should have received as severe a sentence as Thornton.

Actually, Leroy Tozier had made an exceptionally good bargain for both of his clients. In addition to the moderate fifteen-year sentences, which could be reduced to ten years for good behavior, all previous charges were dropped. Tozier suggested the final terms. U.S. District Attorney Harlan accepted them on behalf of the govern-

ment because, it was said, not having to prosecute would
save thousands of dollars in securing witnesses and paying juries and would not put the other work of the court
back two or three weeks. Although these were compelling
arguments, no one ever accepted this official version at
face value. Without exception, Fairbanks was convinced
that the government agreed for a single reason. Every
legal agency would have accepted almost any terms to be
rid of these persistent troublemakers.

The newspapers and most of the town's citizens were
further convinced that Hendrickson and Thornton had
not entered guilty pleas to secure lighter sentences for
themselves and a forgiveness of past transgressions. According to a widely held belief, the prisoners had planned
another escape, but it could not be accomplished while
they were in riveted irons or after they were lodged in the
new jail now nearing completion. Quick action was there-
fore essential, and especially since winter was not far
away. As the *Daily Times* suggested, with steel tanks to
live in ". . . the chances for escape became limited to a
charitably inclined guard handing them the keys and a
ticket to the outside."

The same news story also stated that the first steps toward an eventual escape had already been taken by the
prisoners but was fortunately detected. According to this
report, the prisoners, within two days of having their riveted irons removed, had manufactured a key that would
unlock their Oregon boots. Accidental discovery of the
key was all that prevented an escape even before this
story was printed, said the *Times*. This might have
seemed to be the most preposterous of claims but for the
events shortly to follow.

The fall of 1906 provided an unbelievable contrast to
the previous year, when winter arrived so early and with
such extreme temperatures. On streets, in saloons, wher-

ever people gathered, they congratulated one another on their good fortune. The prolonged, colorful days of Indian summer lingered on into mid-October.

With unusually heavy river traffic that fall, it had been difficult to secure transportation for the prisoners destined for McNeils Island. The first available booking was on September 24 aboard the *Lavelle Young*, three weeks after Hendrickson's and Thornton's guilty pleas.

This three-week interval was a period of ceaseless vigilance for George Dreibelbis. At odd hours day and night, he could be observed checking the cells, the prisoners' irons, the performance of the guards. Nothing was left to chance. There were no routine procedures since the most ordinary daily practices were changed frequently, and above all there were Dreibelbis's unscheduled appearances at all hours.

One day he was present when a reporter for the *Daily Times* was permitted to interview Hendrickson. This was not Dreibelbis's usual custom, and the reporter, obviously embarrassed by his presence, was having a difficult time with his questioning until Hendrickson came to his assistance. He cheerfully explained to the reporter all the precautions taken for his confinement. Then, turning to Dreibelbis, he said in words the reporter faithfully recorded, " 'I want you to understand, Mr. Dreibelbis, that it is not these irons that keeps me in here' and he pointed to the big Oregon boot fastened to his leg—'but the four guards outside here. It would mean death to pass them. So far as the boot is concerned, I can take it off in half an hour.' "

Marshal Perry was maintaining a low profile, counting each day that passed without incident as a blessing. He had been spared recent personal attack and hoped that the newspapers would continue to focus their attention where it belonged, on Hendrickson and Thornton. It was too much to expect.

The *Daily Times* published one of its most scorching editorials upon learning that two of the four men selected to guard the prisoners on their trip were the marshal's personal friends. After stressing the necessity for the utmost security, the editorial proceeded to castigate Perry for his selection:

> The two guards, Darlington and Noon, are not men of experience in handling prisoners. The fact that they were employed as guards was not due to such experience, or because of any ability in that line, but solely because they wanted to get outside without paying their way and had sufficient pull with the marshal's office to accomplish it.

In hiring Darlington and Noon, Marshal Perry had made one of his customary personal decisions. Now he had to appoint one of his deputies to head the five-man party that would escort the prisoners to McNeils Island. He was obviously embarrassed when he called George Dreibelbis to his office to offer him the job. Dreibelbis had long ago decided that it was not a job he wanted; his decision had nothing to do with Marshal Perry's recent action. In the end Frank Wiseman accepted because he had family business requiring attention in Seattle.

Despite prophecies to the contrary, September 24 arrived without incident. That morning, long before any of the regular passengers reached the dock, the prisoners were secured in the central stateroom reserved for them. The day had started for them with an early breakfast, after which they were forced to undress and scrub vigorously under the careful scrutiny of the guards. After this they were made to walk naked into an adjoining room, where they were clothed in new underwear and outer apparel.

Few spectators were on hand when the prisoners were about to be led aboard the *Lavelle Young*. George Dreibelbis and one guard had left the jail ahead of the main party.

Lavelle Young at Fairbanks, 1904

Mackay Collection, Archives, University of Alaska, Fairbanks

Now they were posted, rifles in hand, on the upper deck of the steamer, commanding an unobstructed view of the entire dock, the nearby waterfront, and the nearest stretch of Cushman Street. The prisoners were shuffling down that street in their Oregon boots, with armed deputies and guards on either side. Hendrickson and Thornton received scant attention from Dreibelbis. Frank Wiseman, marching slightly behind the advancing party, could take care of them.

Dreibelbis's attention was focused on the small knots of people scattered about the dock. He knew most of them. There were the usual idly curious, as well as several reporters, who represented perhaps the biggest single contingent of spectators. As Dreibelbis's gaze moved restlessly from one end of the dock to the other, he also recognized friends of Hendrickson who had visited him at the jail. There were not many; there were none of Thornton's. Dreibelbis was momentarily surprised to see Leroy Tozier, one of the last persons he would have expected at this time and place.

Dreibelbis's gaze kept returning curiously to one group of six people whom he recognized as sporting types. He could only make out the identity of three of them because the group kept moving, always in a tight knot, their backs turned toward him.

Now the prisoners had reached the dock and Hendrickson was nearing the gangplank. Dreibelbis's gaze shifted momentarily to the other end of the dock, where two apparently drunken celebrators were attempting to throw a third into the river, to accompanying shrieks of protest and encouragement.

When Dreibelbis's eyes swung back to Hendrickson, it was just in time to see the group he had been watching previously edge closer to the file of prisoners. One of the six seemed to slip or stagger, and in so doing pushed aside one of the guards. At the same instant the group separated and a slender figure dashed up to Hendrickson, throwing arms passionately around his back. A cap fell off and a profusion of dark auburn hair cascaded down her back. It was a woman, Dreibelbis realized in a moment of shock. He could not have known it before because, like any man, she was wearing faded Levis. Women simply did not dress in that fashion!

Everyone was momentarily startled. Then the guard beside Dreibelbis started to raise his rifle. Dreibelbis knocked it down quickly. The girl had already disappeared among her companions. The guard behind Hendrickson jammed a rifle barrel into his back to urge him up the gangplank and onto the ship. The entire incident had taken but seconds and was missed completely by many of the spectators.

Dreibelbis did not visit the stateroom to which the prisoners were led. There was nothing to be said between himself and Hendrickson that had not already been said. Slowly he descended the gangplank and walked to the far end of the dock. There he remained until finally the shrill whistle of the steamer signaled its disappearance around the bend.

13

RIVER VOYAGE

In addition to Hendrickson and Thornton, there were two other prisoners in the party bound for McNeils Island. One was a minor criminal by the name of Kunz, sentenced to two years for forgery. The other was a more imaginative culprit named Bobby Miller, convicted of an express robbery of gold bricks.

Every precaution was taken on the river voyage for the restraint of the prisoners, if not for their convenience. Much more care was given to their accommodations than to those of the paying passengers. Despite the large passenger list on this, the last regularly scheduled sailing of the year from Fairbanks, a double stateroom with three berths, one above another on two sides of the cabin, had been reserved for the prisoners.

On the day prior to the sailing, the prisoners' stateroom was thoroughly searched by Marshal Wiseman, with such meticulous care that even the wooden rod supporting a thin brass tube from which a window curtain was suspended had been removed, along with the curtain. The brass tube, presenting no possible threat, was left in place. Following the search of the room itself, the sole window of the cabin was barred on the outside with iron, and the door handle was removed so the door could be opened only by inserting a key in the lock. Hendrickson and Thornton were further restrained by the Oregon boots, which were never to be removed. Kunz and Miller, not considered dangerous, were not ironed.

In addition to these elaborate precautions, five men had been assigned to guard the four prisoners. Besides Marshal Wiseman there were two day guards, Jack Sowerby and Charles Webb, plus the two night guards accused of political cronyism by the *Daily Times,* James Darlington and Jack Noon. Marshal Wiseman always oversaw personally the feeding of the prisoners. This was done by handing their meals to them on plates with only a single spoon for each man to eat with. The marshal's frequent and unpredictable appearances at all hours was calculated to keep the guards alert. Of the two guards always on duty, one could be seen by curious fellow passengers seated with a rifle across his knees in front of the stateroom door; the other, with rifle in the crook of an arm, paced back and forth on deck before the barred window. While the prisoners were thus well provided for in these particulars, they were not exactly first-class passengers. The only luxuries allowed them were cards, pipes, and tobacco.

If it was a subdued atmosphere within the heavily guarded stateroom, a holiday mood prevailed among the other passengers. It had been a long, hard, but prosperous season, with gold cascading into town from seemingly endless pay streaks. The diverse list of passengers, from miners and merchants to saloonkeepers and working stiffs, had much in common. All had money in their pockets, all had shared the stampede trails. They were companions, and all were determined to enjoy every moment of this long-awaited day. For many it was their first trip to the Outside since being lured north by the call of gold in the first stampedes. For everyone this was going to be a trip to remember—if not precisely in the way they had planned.

The perfect Indian summer weather added to the enjoyment of those first days. The hot, brilliant hours of sunshine were an unexpected bonus so late in the season. One could walk the decks in shirt sleeves when the sun was high. Only the sudden sundown chill, the freezing nighttime temperatures, and the rapidly shortening hours of daylight warned that winter was nearing.

Fast time was made with the aiding current to Tanana and Fort Gibbon at the mouth of the Tanana River, but there the pace slackened as the *Lavelle Young* turned its prow into the swift current of the Yukon on the run to Dawson. Even so, the steamer made steady progress and arrived at the river town of Rampart on schedule. There they were confronted by a totally unexpected development.

As the *Lavelle Young* nosed into the dock, its whistle screaming, the passengers could see a wildly cheering, waving group of men, some throwing their hats into the air. All had packs on their backs or leaning nearby, and most carried rifles. To those on board it was a familiar picture out of their own past. No sooner was the gangplank lowered than the pushing, boisterous mob swarmed aboard. The captain met them but made no attempt to stop the men, if indeed he could have. He, as the others aboard, quickly recognized the unmistakable fever of another stampede. Even as the newcomers were streaming aboard, some were calling back and forth to friends among the passengers who were lining the rails.

The *Lavelle Young* was a small vessel with a passenger capacity of about 150 people. This limit had been stretched when the boat departed Fairbanks, to say noth-

ing of the additional passengers taken aboard at Tanana.
Now it was strained far beyond its designed capacity.
None of the passengers, however, would have wanted the
stampeders to be left behind despite inconveniences.
Some doubled up in limited cabin space; others were
packed into a small, narrow stateroom in the stern near
the paddlewheel, so hot that it was rarely ever occupied.
Still others had to spread their bedrolls on deck and sleep
under the stars.

The stampeders were bound for the Chandalar River
and upper reaches of the Koyukuk in the southern fringes
of the Brooks Range. They would be fellow passengers
until Fort Yukon. In the meantime the *Lavelle Young* was
a beehive of excited humanity.

All the passenger staterooms were arranged around a
central dining room, with their windows on the outer
deck. The large dining salon was the only space that
could be used between meals as a lounging and recrea-
tion room. Thus during meals or at almost any hour of the
day or night it was the scene of noisy, jostling, varied
activity. Between frequent rounds of refreshments, men
exchanged news of this latest strike and recollections of
many past stampedes. Tables were at a premium; endless
card games seemed always in progress. For men and
women alike, this was an exhilarating release from the
drudgery of everyday life. Startled husbands could not
believe it when they saw their wives engage in lively
conversation with some of the town's well-known profes-
sional girls; at home they would have crossed the street to
avoid meeting face to face.

Thus far the trip had been an unalloyed pleasure. But
now Rampart Canyon, one of the most spectacular
stretches of the Yukon River, was behind them. Late in
the evening of September 26, the day after leaving Rampart, the *Lavelle Young* entered the flat, monotonous,

seemingly endless miles of the Yukon Flats. Through this great 200-mile-long northern loop of the Yukon, in places seventy miles wide, the river wanders in myriad, constantly shifting channels. This dreary stretch of the Yukon can be a river pilot's nightmare and for passengers a tedious passage of time.

Steady progress was being made against the sluggish current when, at noon the following day, the machinery broke down. Curious passengers were assured that repairs would soon be made. There was no anxiety, and a few hours later the *Lavelle Young* was again underway. Perhaps too much was being asked of the aging machinery with its overtaxed capacity, for just twenty-four hours later the engine blew a cylinder head. As quickly as possible the *Lavelle Young* tied up at a nearby island for further repairs. They were no sooner undertaken when a fumble-fingered member of the crew dropped one of the bearings overboard in water so deep that it could not be recovered. There was nothing to be done now but wait and hope for an uncertain rescue.

It was September 28, late for any boats to still be plying the rivers. But at Fairbanks it had been said that if river conditions remained good, one last steamer might be sent up. This offered a chance, but the situation was precarious. The *Lavelle Young*'s location hugged the Arctic Circle, some fifty miles below Fort Yukon where the eastern-flowing Porcupine River joins the Yukon. The nearest telegraph stations were at Tanana, now far behind them, and at Eagle, still far upriver. Provisions were running low due to the unexpected number of passengers. Slush ice beginning to float by warned that the Chandalar and Porcupine rivers had started to run ice, and the Yukon itself always froze first in these flats. Even the ducks and geese had long ago banded together in enormous flights on their southern migrations from this, the

greatest of North American breeding grounds. Now only
the little brown cranes were still migrating, but even they
were seen in ever-lessening numbers.

The passengers, however, had witnessed far too many
hardships to be unduly disturbed by the present circumstances. They remained cheerful. Unaccountably, someone had a football in his luggage. A rough field was beat-
en out on the sandy island shore, and many of the men
spent hours at the game. Others broke out guns and
hunted, but none with the success of the old Indian who
one day drew alongside in his beaver canoe with a freshly
killed moose. He was quickly persuaded to part with most
of it, and dinner that night was a gala affair.

Three days later, on October 1, passengers and crew
were astonished to see a stern-wheeler appear, as if in a
mirage, from around the bend, hurriedly check its downriver speed, and, to a screeching of whistles on both sides,
slowly maneuver alongside. It was the Northern Commercial Company packet *Ida May*, loaded with passengers for Fairbanks. Everyone was soon crowding the rails
on both boats, shouting greetings as they recognized
friends, yelling questions, exchanging news. It was a
first-class old-time hometown reunion. Everybody
seemed to know everybody.

Jubilation, however, was soon quelled. The *Ida May*
could provide little assistance. The best it was able to do
was tow the hapless *Lavelle Young* a few miles downriver
to a safer anchorage. Then, with more urgency than ever
for speed, so it could report the *Lavelle Young*'s plight
and summon help, it pulled off downriver. The *Lavelle
Young* and its passengers were again left to their isolated
fate.

There was less exhilaration now among the passengers.
Food was becoming scarce. The weather was growing
colder, and the ice ran heavier in the river. The night of

the *Ida May*'s departure, some of the more indomitable
women tried to liven spirits by providing a midnight
lunch of treasured home-baked cookies, sweet breads, jel-
lies, and other delicacies that had been intended for
friends and relatives they were to visit.

Despite such efforts, it became increasingly difficult for
passengers to remain cheerful. It was not possible at their
new location to get ashore conveniently, so the football
games were not resumed. Now the men turned to playing
poker with a kind of desperate urgency, and the stakes
rose progressively.

As their own fate became more uncertain, the passengers evinced increasing sympathy for the prisoners. At
their trial, both Hendrickson and Thornton had appeared
to be weak physically from prolonged confinement. Most
passengers expressed the opinion that the prisoners
should be given some relief from their cramped quarters.
The only one undisturbed by the openly voiced suggestions was Marshal Wiseman. If he heard them, they were
never acknowledged. He continued his rigid surveillance
undisturbed by the criticism that swirled about him.

It was fortunate for everyone aboard that they did not
have to wait for the *Ida May* to reach Tanana before help
could be summoned. The owners had decided to risk dis-
patching the *Seattle No. 3* from Fairbanks to Dawson. If
the ship could make it through, it would be available at
Dawson for next year's early summer rush. If the ice became too heavy, the *Seattle* could be frozen in for the
winter in some slough, safe from the shattering impact of
the spring breakup. It learned en route of the *Lavelle
Young*'s difficulties, and two days after the *Ida May*'s departure it pulled in alongside the *Lavelle Young*.

The arrival of the *Seattle No. 3* brought none of the
elation occasioned by the *Ida May*'s appearance, only relief and the hope that nothing more would go wrong on

this apparently jinxed voyage.

The crew of the *Seattle No. 3*, alerted to the problems of the *Lavelle Young* and knowing that repairs were out of the question, took the disabled ship in tow. Strong cables lashed the two boats together, and then the *Seattle* propelled both ships against the current. Progress was excruciatingly slow, but they were underway again.

At six o'clock the following day they reached Fort Yukon. The stampeders bound for the Koyukuk departed to the accompaniment of rousing cheers and good wishes. Above Fort Yukon the river was free of ice, and the boats were able to make better progress. Circle City was reached on October 5, and soon they were in the hills again, to the relief of all on board. It must have seemed incredible to many that they had now been on this voyage for eleven days: from Circle City across the highlands by the old Circle trail, Fairbanks was only 160 miles distant. But now the journey was rapidly nearing its end.

The captain announced that a stop would be made at Nation City to take on more wood for the ravenous steam boilers of the *Seattle No. 3*. Then there would be only Eagle, almost astride the Alaska-Canada boundary, before Dawson itself was reached. The passengers heaved a collective sigh of relief. Nothing more could go wrong now—could it?

14

NATION CITY

Nowhere on board the *Lavelle Young* had the waiting been harder to endure than in the prisoners' tightly guarded stateroom. Hendrickson and Thornton knew that the success of their long, carefully rehearsed plans would be determined in large part by the weather, and each day that dragged by with the weather turning steadily colder reduced their hopes.

Thornton was turning increasingly surly, a bad sign; his vicious, unreasoning temper could destroy everything. For their plans to have a chance, they needed the cooperation of the other prisoners, Miller and Kunz. Hendrickson had to keep them in good humor while also holding Thornton in line.

The brash, irrepressible Miller had scoffed openly at Hendrickson's first reference to a planned escape. "You're crazy. Look outside at those guards. We're pigeons in a cage."

They were seated at a table, their scant meal just finished, the utensils soon to be collected by one of the guards. Hendrickson made no reply. Slowly he reached for his pipe. Miller thought he was about to light up and pushed a package of tobacco across the table toward him. Hendrickson shook his head. Instead he reached over to pick up a spoon and began carefully scraping out the bowl of his pipe. Miller noticed then that the bowl was so heavily caked that it could not have held a thimbleful of tobacco. He watched Hendrickson's careful movements derisively until a small metallic object fell with a little clink to the table's surface. Miller stared at it incredulously, then started to laugh until the tears came. He picked up the small object, held it in his palm so the suddenly alert Kunz could see it also. It was a tiny jeweler's file.

"I'll be damned. Perhaps you *can* do it. I tell you what. I fooled them once, and it will be a pleasure to help you do it again. Count me in."

"Understand," said Hendrickson, "when we are free it is every man for himself."

"Oh, I'm not going anyplace," replied Miller. "We'll help you pull it off, won't we, Kunz, but I stay right here. I got a light sentence, and the gold is waiting for me to spend when I get out. I'm not adding to my time."

"You're all crazy," said Kunz. "I'll have no part of this."

Thornton swore savagely, and Hendrickson motioned him to be quiet. He said nothing to Kunz. He would leave that to the irrepressible Miller, who could honey talk the halo off an angel.

"Why does it have to be Nation?" Miller asked repeatedly. "You fellows aren't in the best of shape, you know, and it will get a hell of a lot colder before it gets any warmer. You couldn't pay me to go beating the brush this time of year."

Thornton would have grabbed Miller by the throat if Hendrickson had not stepped between them. This peacekeeping was not getting easier. Miller took pleasure in goading Thornton, and Hendrickson found it increasingly difficult to control Thornton's outbursts.

"There is no other way," Hendrickson answered when things had settled down again. "I know the creek, the lay of the land, and it has to be a place where we can get heavy clothes, food, guns. They don't grow on the willows."

"This creek and the mines you mentioned. How far are they off the river?" Miller asked.

"Fourth of July Creek. It enters the Yukon on the south shore, hidden behind Nation Reef. The mines? They are about twelve miles above the mouth."

"And Nation City? Where is it?"

"Rightly speaking, it is not much of a city. Three or four cabins and a wood camp, that's the extent of it. Good, convenient timber for the riverboats, and it's the river landing for the mines on Fourth of July. It's two miles above the mouth of the creek at the nearest decent anchorage."

"You won't be exactly crowded for company, that's a fact," Miller said.

"Better that way," Hendrickson said. "We haven't found people any too friendly lately."

Hendrickson had to wait until the *Lavelle Young* was in tow of the *Seattle No. 3* before putting the first part of his plan into operation. With Marshal Wiseman's twice-a-day inspection of the stateroom, most of the scheme had to wait until almost the last moment, but certain preliminaries could be undertaken.

The brass curtain rod played an important part in Hendrickson's plans, but he could not afford to have its absence noted. The second day out he had removed the rod and set it in a corner. If Wiseman noted its removal, he could always be told that the rod blocked the sole view of the outside world through the cabin window—reason

enough for its removal. The action, however, went unnoticed. Two days later Hendrickson moved the rod to a
carefully selected hiding place. Its absence still went undetected. It was only then that Hendrickson felt safe to
proceed with his next move.

With Miller standing next to the window, cajoling the
guard as he paced back and forth and at the same time
effectively blocking his view, and with both Thornton
and Kunz further shielding his movements, Hendrickson
set to work with his jeweler's file and the brass curtain
rod. First he sawed off an eight-inch section of the rod,
divided and flattened it into two pieces. From one of the
pieces he painstakingly fashioned a small saw. It was
slow, precision-type work that required several sessions.
More than once he was interrupted by the surprise arrival
of Marshal Wiseman. But so meticulously had every
movement been rehearsed that, warned by Miller, everything was thoroughly concealed before the stateroom
door opened.

After completing the saw, Hendrickson made a key
from the other section of rod with which to unlock the
Oregon boots. When at last he finished the job, Miller
called Kunz to take his place at the window. He walked
over to inspect the key.

"I don't believe it," Miller said admiringly. "It's beauti-
ful, but will it work?"

Hendrickson bent over and inserted the key into the
lock. A gentle turn and the boots dropped off with a dull
thud. Then calmly he replaced the boots, rose, and
grinned at the nonplussed Miller.

Hendrickson needed one other tool. It was already at
hand, one of the two brass hooks that had supported the
curtain rod. It could be unscrewed for use whenever
needed.

Hendrickson waited for the *Lavelle Young* to clear Fort

Yukon before proceeding with his final plans. By this time Kunz had become a willing if not enthusiastic conspirator. Also, in the intervening days, Miller had come to be on familiar terms with the guards, particularly Darlington and Noon. This was not difficult to accomplish since the trip had become almost as monotonous for the guards as for the prisoners. Often Miller would hold the patrolling guard in lengthy conversation if Marshal Wiseman was known to be safely out of sight and hearing.

Shielded from sight of the guard by Bobby Miller's connivance, Hendrickson and Thornton took turns lying on their back on an upper berth, cutting an escape hatch. First the brass hook was used to make several adjoining holes in the half-inch ceiling, which also constituted the roof of the steamer. When the holes were wide enough to provide an opening for the saw, it was used to lengthen the groove. This had to be repeated on each of four sides and, even with both men working, it was slow business. When they were too tired to work longer, both the holes and the incisions made by the saw were filled with white soap, which being the same color as the ceiling effectively covered up all evidence of their work. The holes went undetected by Marshal Wiseman during his daily inspections. At last a section fifteen inches long and ten inches wide was sawed out of the ceiling and fastened in place until the opportune time for escape.

About 5:30 P.M. on October 6 the *Seattle No. 3* with the *Lavelle Young* in tow pulled alonside the landing at Nation City. This late in the season it was completely dark by that hour and, as Bobby Miller said, even on a trip as beset by trouble as this one had been, not all one's luck could be bad. Certainly nothing could have aided Hendrickson and Thornton's plans more than darkness.

The powerful searchlights of both ships were turned on the shore to aid the crew in its hurried loading of cordwood for the *Seattle*'s boilers. On board ship everyone was preparing for the six-o'clock dinner.

Marshal Wiseman had seen to the prisoners' feeding fifteen minutes before the vessels docked. First he had given the stateroom his customary inspection; the dimly lit interior helped obscure the work of the prisoners. Then he ordered the night guards, who had just come on duty, to pass in the prisoners' food. Contrary to his usual habit, he waited, his back turned to the stateroom door, until the prisoners finished. Miller and Kunz ate sparingly, but Hendrickson and Thornton wolfed down their helpings and everything that the other prisoners left. In an incredibly short time their dinner was finished. Marshal Wiseman gathered up the empty plates and utensils and, with a nod to the prisoners, backed carefully out the door. He reached the dining salon just as the other passengers were being served.

As soon as Marshal Wiseman left the prisoners' stateroom, Miller and Kunz pulled chairs up to the table and began a rowdy game of cards. James Darlington, just come on duty, paced leisurely back and forth along the deck, pausing occasionally to stretch and yawn.

Thornton, meanwhile, waited until he was sure Marshal Wiseman was out of hearing. He then went to the window, calling out to Jack Noon that he needed matches. Noon rose cursing from the comfort of his chair propped against the stateroom door. He rummaged through his pockets until he found spare matches, strolled over to the window, and handed them to Thornton.

Thornton's action had been intended simply as another diversion while Hendrickson wriggled his way to safety through the hole in the ceiling. Instead it almost ended in disaster. Thornton's Oregon boots had already been unlocked in preparation for his own escape, and now as he faced Noon through the barred window they accidentally fell from his legs with a terrifying clatter.

"What's going on in there?" Noon demanded, instantly alerted by the racket.

The nimble-witted Miller saved the situation. "Look out, you clumsy oaf," he yelled at Thornton. "See what you've done now." He pushed Thornton from the window, shaking his pant's leg for Noon to see, and swearing. "If you had to knock over that slop bucket, why couldn't you kick it in your own direction?"

Noon, laughing uproariously at Miller's apparent discomfort, turned away from the window and returned to his chair. Thornton did not need a second reprieve. Quickly he followed Hendrickson's exit through the ceiling.

Hendrickson had not waited for Thornton. Scrambling through the opening to the upper deck, he found himself near a ladder leading down to the deck on which their stateroom door opened, now guarded by Noon and Darlington. Waiting until Darlington reached the end of his normal patrol and had turned his steps in the opposite direction, Hendrickson slid down the ladder, turned a corner and, with a running jump, cleared the five feet to the deck of the *Seattle No. 3*. A moment later he was over the side into the protecting cover of the brush growing close to the riverbank.

The escape might have gone completely unnoticed until Eagle or Dawson, as the prisoners were counting upon, except for Thornton's second miscalculation. Not seeing Darlington but assuming that he had probably stopped in his rounds to talk with Noon, he too scampered down the ladder. Darlington, however, hoping to catch another glimpse of a pretty pair of ankles he had noticed before, had extended his normal patrol as far as the dining salon. Disappointed in his quest, he turned to retrace his steps and, just rounding the corner that had protected Hendrickson, met Thornton face-to-face. Thornton never paused for a second. The next instant he was around the corner and, following in Hendrickson's steps, was soon ashore.

Darlington was dumbfounded. He could not credit his senses. He continued his pacing, his mind refusing to believe what he had seen. It was not until he reached the prisoners' cabin that comprehension dawned. He cast a hurried look inside the room. Seeing neither Hendrickson nor Thornton, he sprinted for the dining salon, forgetting completely Wiseman's instructions to fire his rifle twice in event of any emergency.

Dinner had been in progress about fifteen minutes when Darlington burst into the salon, flourishing his rifle, and rushed up to Marshal Wiseman's table.

"They are gone," he yelled. "Hendrickson and Thornton have escaped."

Marshal Wiseman exploded out of his chair and with Darlington trailing, rushed to the prisoners' cabin. The half-dressed day guards, awakened by the yelling and commotion, were just behind them, shirttails trailing from half-buttoned trousers. The marshal was the first man into the cabin. The gaping hole in the ceiling was all the confirmation of the escape needed.

"You will get ten extra years for your part in this," Wiseman yelled at the remaining prisoners.

"They threatened to kill us," Miller pleaded. "Thornton held this saw to my throat. You can see the marks."

This was no time to be trading words with Miller. Wiseman turned on the two night guards. "Where did they go?" he shouted.

"I don't know," Darlington admitted.

"You saw him and you don't know?" Wiseman's usually low-keyed voice roared in anger. "You're fired. You two incompetent idiots can swim to Dawson for all of me."

"Sowerby," the marshal shouted to one of the day guards while gesturing toward the remaining prisoners, "you take charge here."

By this time the deck was becoming so crowded with male passengers that they threatened to push one another over the rail.

"I'm deputizing every able-bodied man of you," Wiseman roared at the crowd.

To the other day guard standing at his shoulder he directed, "Charlie, take a dozen of these men, split up into groups, and search every nook and cranny of these boats. Don't you miss an inch."

Turning to the remainder of the eager crowd, he ordered, "Get warm clothes and meet me at the gangplank in ten minutes. We will find those fellows if it takes us until Christmas."

"Oh, no you won't," bellowed a barrel-chested giant of a man who pushed his way through the crowd as if it were so much chaff. "I'm the captain here. When the cordwood is loaded, we move."

Marshal Wiseman protested. "We have two dangerous criminals out there and a lot of innocent, unsuspecting people."

"I wouldn't care if it were Jack the Ripper. I won't risk my boat and passengers on the river at this time of year for one extra hour."

In the end a compromise was reached. The captain agreed to hold the boat until midnight but not one minute longer. The shore party fanned out under Wiseman's direction to scour the riverbank, with strict orders to be back promptly by eleven thirty. They presented a weird sight from the deck of the ship, lanterns and candles casting tiny firefly flickerings of light. One of the ship's searchlights continued to assist the crew until the last of the cordwood was loaded. Then it joined in intermittent sweeps of the riverbank. The lights' glare in the pitch darkness was like flashes of lightning—and about as useless to the searchers.

Long before the agreed-upon hour of eleven thirty, the searchers began straggling back aboard the ship. Many of them were shivering and chattering with the cold. Two of them had fallen down the steep bank into the river, to be rescued by their companions. In the near-zero nighttime temperature, the initial enthusiasm of the chase quickly evaporated. Some were heard to remark that if the prisoners were willing to take this kind of misery, they deserved to be free.

The search of the two boats of course turned up no trace of the prisoners. The chase ashore, after discovery of some footprints where the prisoners first hit the riverbank, was equally futile. Even Marshal Wiseman had to admit that further search under the circumstances was useless.

There were still other things, however, that needed the marshal's attention. First he made the hard decision to proceed with the boat to Eagle, thirty-two miles upriver, where the Fairbanks authorities could be notified by telegraph of the latest developments. The search for his vanished prisoners, however, could not await his return. He deputized two of the residents of Nation City to launch the search in his absence and, on his own responsibility, offered a reward of $250 each for the apprehension of the ex-prisoners. None of the usual strings of "arrest and conviction" were attached to the reward. It would go to whoever marched the two convicts into camp. If Marshal Perry should welch on the reward and it had to be paid out of his own pocket, so be it.

In the little time still remaining to him, Marshal Wiseman arranged to have another man from Nation City dispatched over the trail by lantern light to warn the miners on Fourth of July Creek of the escape. Another, an Indian, was started downriver by canoe to warn an Indian camp fifty miles below, then to proceed to Circle City with warnings for the roadhouse men there, as well as for anyone else encountered along the river.

He had done what he could. Promptly at midnight, with a prolonged blast of its whistle, the *Seattle No. 3* cast off for the run to Eagle.

15

PURSUIT

For six interminable, shivering hours Hendrickson and Thornton hid out in the heavy brush near the riverbank while the posse beat back and forth in vain pursuit. At times some of the searchers passed within a few feet of their hiding place. Only the unfamiliarity of most of the searchers with such a hunt, plus their discomfort and quickly waning enthusiasm, may have prevented discovery.

By the time Hendrickson and Thornton heard the lonely blast of the departing steamer, they were so stiff from the cold and dampness they could barely stand. The light clothing Wiseman had compelled them to wear provided no protection against the zero temperature. When they were able to move, they headed by unspoken agreement for the roadhouse, knowing their only hope lay in acquiring warm clothing. Even food and rifles were of secondary importance.

Kerosene lamps glowed dimly through the low framed windows of the roadhouse when Hendrickson and Thornton staggered the last few steps to their destination. They rested briefly, regaining their breath, before deciding on their next move. But even as they waited, the roadhouse door opened and three men stepped out into the darkness. Each was heavily armed, unusual in a land where men rarely carried weapons. They had obviously not finished the discussion they had been holding inside the roadhouse, and their voices carried easily to where Hendrickson and Thornton were hiding a few feet away. This was how the escaped prisoners learned that Marshal Wiseman had deputized two men to head a search for them in his absence, and of the reward offered for their capture. It was shattering news. They knew that the offer of the reward made them marked men. Nation City had become an armed camp. Gone was any hope of surprise.

They had one remaining hope, and that a forlorn one. Some ten to twelve miles up Fourth of July Creek was the tiny settlement of Montauk. It consisted of a roadhouse and six log cabins occupied in winter by men who worked on the nearby placer claims. Residents of Montauk might not yet have been alerted to their escape. If the fugitives' strength held out long enough, they might make it to the roadhouse. They struck out in pitch darkness along the trail leading up the creek, stumbling on the uneven going, stopping often to rest. Hendrickson, since his capture on Birch Creek, had spent nearly a year in jail, Thornton almost as long. Their lengthy confinement and lack of exercise had weakened both men. Sunrise found them still short of their destination.

As they struggled ahead, sick from the effects of the cold, even the gradual warming from the sun provided no relief. Thornton began grabbing big handfuls of frozen blueberries that grew profusely alongside the trail, gulping them greedily. Despite Hendrickson's warnings of the effect on a cramped and empty stomach, Thornton would not stop. As they neared the roadhouse he became violently sick, falling in the middle of the trail, retching uncontrollably. Half an hour was wasted until he could travel again.

When Montauk was finally reached, it was apparent from a careful survey that none of the cabins were occupied and that the only residents of the camp must still be inside the roadhouse. While Hendrickson had waited for Thornton to recover sufficiently to travel, he had fashioned two heavy clubs from dead spruce limbs. Now he handed one of them to Thornton and told him to follow. When they burst through the roadhouse door moments later, they presented such a wild, desperate appearance to its four lone occupants that none of them thought for a moment of offering any resistance.

Hendrickson found ropes and bound three of the men. The fourth, the roadhouse keeper, was ordered to start cooking and not to stop until he was told. Meanwhile both men fell like hungry wolves on the remnants of the unfinished meals sitting on the kitchen table. Hendrickson thought he had lost the ability to feel more cold, but as the cabin warmth began to seep into his bones, he started to shiver so convulsively that the spoon he was holding fell clattering to the floor.

Hendrickson and Thornton ate until they thought they could never eat again, then still continued to lunch throughout the day. After their first big meal their bodies cried for sleep, but before they could allow themselves to think of that, they had to find warm, heavy clothing and guns. They scoured the roadhouse and cabins, finding all the clothing they needed but not a single gun. What they did not know, and certainly were not told, was that one of their pursuers had arrived from Nation before them, and that all guns had been collected and carefully hidden. No other precautions were taken in the belief that the escaped prisoners could never reach the roadhouse, or probably even knew of its existence.

The strange absence of any guns was a rankling source of worry to Hendrickson. Several times he questioned roadhouse keeper T. B. Villineve about it, but the man was so cooperative in every respect that Hendrickson had to conclude that it was, after all, an implausible coincidence. As for Villineve, he could afford to be cooperative, knowing that posses were already scouring the country in search of the escaped prisoners and would sooner or later conclude that they had headed for the roadhouse.

Throughout that Sunday, the day of their arrival, Hendrickson and Thornton remained at the Montauk roadhouse, alternately eating, sleeping, resting, and recovering their strength. Hendrickson knew as surely as Villineve that they would not be safe at the roadhouse for long. Safety lay in keeping as wide a distance as possible between themselves and their pursuers. They were still too weak, however, for hard travel, and Hendrickson decided reluctantly in favor of an overnight rest at the roadhouse and an early start the following morning. Having made this decision, he set to work with Thornton's help to make up trail packs containing food, blankets, tarps, spare clothing, and everything else they would need for a long journey.

One of the hardest decisions Frank Wiseman ever had to make was to proceed with the *Lavelle Young* to Eagle. He felt compelled, however, to advise Marshal Perry of the prisoners' escape. Moreover, a quick check of the resources at Nation revealed a woeful lack of supplies. He knew that the search for Hendrickson and Thornton could be a long one and that, once started, must be followed to the end. Eagle, the only available source of supplies he would need for the winter trail, clearly left him no alternative but to proceed there.

Wiseman deputized two men to lead the search in his

3-170-223

Seatle "No. 3" at Eagle, Alaska

Eddy Davis Collection, Archives, University of Alaska, Fairbanks

absence: a Captain Donovan and William Gertz, a tough
Montana ex-sheriff. Knowing the job was in these capable
hands, and with the reward as an added incentive, Wiseman felt somewhat better. However, these preparations in
no way lessened his determination to return with the
least possible delay.

The *Lavelle Young* in tow of the *Seattle No. 3* reached
Eagle early the following morning. Guards Darlington
and Noon were fired on the spot. The prisoners Kunz and
Miller were sent on ahead in the custody of the two remaining Fairbanks guards, and two additional guards
who had been hired at Nation. The party was ordered to
proceed without delay to Dawson, Whitehorse, and then
Skagway, where they were to board the ocean steamer for
Seattle. Marshal Wiseman did not expect to see the party
again.

All of these arrangements were concurred in by Marshal Perry, along with Wiseman's decision to return to
Nation and personally conduct the search for Hendrickson and Thornton. While these details were being
arranged, the supplies Wiseman had ordered were assembled, a canoe was found, and two men were engaged
to accompany Wiseman on his return to Nation City. He
arrived at the Nation roadhouse late that night and fell
almost immediately into an exhausted sleep. He had not
closed his eyes since the night prior to the escape, over
thirty-six hours earlier.

In Marshal Wiseman's absence the posses had been
busy. Men had split up into two groups and scoured the
riverbank both above and below Nation City. The broad
avenue of the river afforded the most logical means of
escape, and so it was the first possibility explored. Despite the daylong search, no trace of the missing men was
found. So far as could be determined, no boats were miss-

ing from their customary moorings. That appeared to narrow the search to trails radiating from Nation City. Of
these there was only one of importance: the trail to Montauk. No one wise to the wilderness, and in the weakened
condition of Hendrickson and Thornton, would think
twice of choosing the trackless stretches of tundra and
muskeg bordering the Yukon.

From this point the exact sequence of events becomes
somewhat confused. Contemporary accounts are less
concerned with the details of what happened than with
its exciting highlights and dramatic conclusion. It is
known, however, that after spending the night at the
Montauk roadhouse, Hendrickson and Thornton set out
early the following Monday morning on the little used
overland trail to Seventymile. It does not seem to have
been a wise decision, and one can only guess at their
reason. It was probably their intention to find temporary
refuge across the international boundary in Canada and
somewhere along the way to acquire the guns indispensable to their survival.

Throughout much of the day Hendrickson and
Thornton struggled ahead on a steadily deteriorating
trail. Rarely used in recent years, it had become more of a
drainage ditch through the surrounding tundra than a
trail. Moose had used it more than men of late, punching
holes through its spongy surface. Later, when it froze
harder, traveling would be easier, but now the partially
frozen surface supported the men's weight only part of
the time, so they were alternately walking on a level surface, then plunging with their next step into a sucking
quagmire. Hendrickson and Thornton had underesti-
mated the difficulty of the trail and had overestimated
their own strength. By midafternoon they knew they
would have to retrace their steps and take their chances
on the river route. If their pursuers had not yet reached

the roadhouse, they might even find another night's rest there before pushing on the next morning.

Strength ebbed from legs too long unaccustomed to such demands and, as the men grew more tired, cold gripped them harder. Despite their warm clothing they started to shiver again. They had not dared start a fire that might attract attention, and the cold food they ate seemed to chill them further. Each hour their progress became slower. It was long after dark when they topped a slight hill and saw the distant lights of the Montauk roadhouse. At that point they left the trail and took a detour through the brush to the cabin.

While Thornton remained hidden in the shadows, Hendrickson crept silently to a lighted window and took a cautious look inside. Sitting comfortably at the kitchen table, coffee cup in hand, was Marshal Wiseman. Two of the other men at the table he recognized as the men he had seen that night outside the Nation roadhouse. Another, a stranger, sat with a rifle across his knees where he commanded an unrestricted view of the outer door. They were begging for an unannounced visit.

16

JOURNEY'S END

The day Hendrickson and Thornton spent on the Seventymile Trail, with such disheartening results, was more successful for Marshal Wiseman. He started out early that morning with a posse to examine the trails leading from Nation City. It was not an easy task. The trails had been tramped back and forth to such an extent since the escape that earlier tracks were often obliterated. It was noon before the posse arrived at the Montauk roadhouse and confirmed their growing certainty that this had been the route taken by Hendrickson and Thornton.

The roadhouse residents had not ventured beyond the immediate premises. They had no idea of Hendrickson's and Thornton's plans, only the wish never to lay eyes on either of them again. Consequently they were of slight help to the posse on its arrival.

Wiseman did not know in what direction to push the search. Neither of the two prime possibilities—the trail leading up the creek to the mines and the Seventymile Trail—showed evidence of recent travel. Hendrickson and Thornton had employed the same subterfuge upon leaving the roadhouse that morning that they used on their return. They had struck out through the brush to hide their tracks and did not join the Seventymile Trail until putting the roadhouse three miles behind them.

In the end, Marshal Wiseman took the trail to the mines. The man sent from Nation to warn the miners had gone no farther than Montauk. Even if it should turn out that his prisoners had taken another direction, the miners must be alerted to their danger.

The posse arrived back at Montauk, tired and discouraged, just as Villineve was preparing dinner. For Wiseman it had been a long, frustrating afternoon. No trace of the prisoners had been found. His only consolation lay in the hope that his slow process of elimination was gradually drawing the net tighter. But where now? The Seventymile Trail remained, but he had already determined that it showed no sign of travel. While the other members of the posse relaxed awaiting dinner, Wiseman paced restlessly in the cabin. At last, grabbing coat and hat and telling Villineve to hold his dinner, he left.

Wiseman had no conscious plan. His steps took him by some kind of gravitational pull to the Seventymile Trail. Walking was better than staying cooped up inside the roadhouse, and slowly his irritation ebbed away. Dusk was starting to fall when he saw a small hill up ahead on the trail and decided to go that far before retracing his steps. It was so dark when he topped the slight rise that he almost failed to see the footprints. He dropped to his knees to make certain. Sure enough, there was the fresh trail he had been hunting for so long. Somewhere up ahead was his quarry. His plans for the morrow were suddenly clear.

The posse hit the trail the following morning as soon as it was light enough to travel. Well rested, they covered the miles quickly and soon reached the hill where Wiseman had made his discovery the previous evening. But now they stared disbelievingly at the trail. What they saw was not one but two sets of tracks. One stretched up ahead as Wiseman had seen them the night before, the other set marked a return. Quick examination showed where the more recent tracks broke off the trail, heading back toward Montauk. Wiseman cursed this loss of precious time. Leaving two of the posse with the slower job of following the prisoners' trail through the brush, Wiseman and his other deputy headed directly back down the Seventymile Trail.

Hendrickson and Thornton took a circuitous route past the roadhouse and made camp scarcely a mile beyond. It was impossible for them to go farther. They were now close to the Nation Trail they would be following the next morning, but carefully shielded from it by a heavy stand of timber. Again they had to make do with a cold meal and without the comfort of a fire. Despite the fresh-cut spruce boughs spread beneath their bedrolls, the dampness of the ground seeped into the very marrow of their bones. When they awoke at daybreak to take up the trail again, every movement was slow and painful. Their only chance of escape now lay in reaching the river by the shortest possible route ahead of the fast-closing posse.

Throughout the afternoon of the previous day, Thornton had grown increasingly morose. Hendrickson could read the signs of trouble, but hoped that a night's rest would revive Thornton's spirits. The rest, however, did nothing to improve his temper. Thornton brooded silently throughout their quickly prepared breakfast, and when making up his pack would have left half the gear behind but for Hendrickson's sharp warning.

After their first hour of travel, Thornton started to lag behind. Hendrickson urged him to take over the lead, hoping it would encourage him to speed up his pace, but

Thornton refused. They continued their slow progress, with Thornton falling farther behind on each bend of the trail. Hendrickson waited at the end of one particularly long, straight stretch for Thornton to come in view. Minutes went by with no sign of him. Disgustedly, Hendrickson eased off his pack and left it beside the trail while he headed back to see what had happened. He found Thornton sitting on a dead stump.

"I'm going back to the roadhouse," Thornton announced before Hendrickson could say a word.

For a time Hendrickson just stared at him in dumbfounded disbelief. "You know what that means," he said at last.

"Hell yes, I know. I'll be warm."

Hendrickson recognized the futility of argument. Besides, he could travel faster alone. Without another word he turned and started back toward the river.

Thornton remained without moving for some time after Hendrickson disappeared from sight. Then, taking his time, he headed back in the direction of the roadhouse. He had almost reached it when he heard voices up ahead on the trail. The last thing he wanted was to encounter members of the posse who might shoot before asking questions.

The trail ran close to the narrow creek. Thornton splashed into its frigid water and up the far bank, heading for cover of the brush and the roadhouse. Frank Wiseman had him in his rifle sights as he clambered up the bank. He did not pull the trigger. He figured that he could outrun Thornton in a footrace, but fear gave Thornton added speed. He reached the roadhouse first, burst through the door, and rushed up to an ashen-faced Villineve. Thornton threw his hands in the air. Villineve ducked as though to avoid an impending blow.

"Arrest me. I'm your prisoner. Arrest me."

Yukon natives in birch bark canoe

Charles Bunnell Collection, Archives, University of Alaska, Fairbanks

Wiseman was only seconds behind. He pulled Thornton's upraised hands down and behind him and snapped cuffs over his wrists.

Villineve had to wait awhile for his reward, but in the end he did receive it. It was probably the easiest earned reward in manhunt history.

Hendrickson's troubles would not be over on reaching the river. He would still have to find a boat before making his escape. Not only that, but the reward offer would have made a hunter out of every man in Nation City.

As Hendrickson struggled to cover the remaining miles to the river, he had still other worries. He was woefully short of supplies. He wished he could have taken a part of Thornton's, but that was out of the question; he was barely making it with what he now carried. Then there was his lack of a gun, which might be the deciding factor. These, however, were worries that could take their place in line. His immediate problem was finding a boat.

The miles grew longer. With each step he found himself leaning more heavily for support on the club he had fashioned for such a different purpose two days before. His first glimpse of the river, framed through a golden shower of falling birch leaves, seemed like some phantom scene.

Hendrickson would have preferred to avoid Nation City altogether, but where else could he hope to find a boat? He approached the riverbank cautiously, keeping inside the shielding border of trees. He spotted a boat finally at a considerable distance from cover. Running, crouched low to the ground, he reached the boat only to find it chained and padlocked to a huge drift log. The next boat he found had no oars. Others he spotted were too close to cabins. He turned his steps back downstream toward Nation Reef. He had proceeded a short distance when he saw a boat in

the slack in-shore current, its occupant apparently heading for a wide gravel bar less than a hundred feet away.

As he watched, the boat ran ashore and the man
stepped out. He did not carry a gun. Things were working
out. The man could be forced to run him across the river.
He would be left there unnoticed and unmissed, perhaps
for hours. By that time Hendrickson would be many river
miles away, and the river left no tracks.

Everything would depend on the next few seconds. The
man approached closer. He was bent over, scanning the
ground intently. At length he hunched his shoulders and
turned back in the direction of the boat. At that same
instant Hendrickson broke from his cover with a chilling
scream, waving his heavy club above his head. The surprise was complete. The man stumbled, then sprawled on
the sandy shore. Hendrickson loomed menacingly above
him, his club still raised.

"You don't need that," the man said. "I'll give no trouble. Figured you'd still be around."

Hendrickson motioned the man to rise. "Move easy.
Get in the boat and row me across."

"Sure. Just as you say."

The man stepped into the boat and sat down on the
thwart facing Hendrickson. "You push off," he said.

Hendrickson bent over to give the boat a shove. It took
barely an instant, but when he looked up to face the man
at the oars, he was staring straight down the double bar-
rels of a shotgun.

On just such a slight miscalculation often rests the fate
of men. One instant Hendrickson had both the boat and
the gun he needed, had he but known it. The next instant
he had lost everything.

Wiseman left Thornton at the Montauk roadhouse in
charge of the deputy who had accompanied him that

morning, with instructions to await the arrival of the other members of the posse. All of them were then to proceed, with Thornton in tow, to Nation City. Hendrickson's destination was now known. Wiseman set out immediately to track him down. He could have spared himself the trouble. When he reached the Nation roadhouse, William Gertz already had Hendrickson safely in custody. The manhunt was ended.

It was now six o'clock in the evening and the river could not be tackled in the darkness. Wiseman had to content himself with preparations for the next day. No powerboats were available. The *Lavelle Young* and the *Seattle No. 3* had been the last scheduled upriver runs for the season. There was no alternative but to pole the tough thirty-two miles upstream to Eagle. A boat was found, two experienced rivermen hired, and all other preparations completed for the next morning's start.

The party cast off at dawn, Hendrickson and Thornton securely ironed together and chained hand and foot. They were still within sight of the roadhouse when they noticed smoke around a bend of the river below them. They waited and soon the *Isom*, largest of the Yukon River steamers, appeared and pulled into the Nation landing for wood. The *Isom* had run solidly aground on a bar in the Yukon Flats earlier, where its owners figured it would have to be abandoned. Miraculously, at this low stage of the river, it had been worked free. Wiseman lost no time transferring his prisoners to the much safer, more comfortable accommodations of the *Isom*.

On October 11 Wiseman and his prisoners arrived at Dawson, where the Canadian Northwest Mounted Police allowed him to lodge Hendrickson and Thornton in their jail overnight. They even assisted in filling the prisoners' leg-iron locks with hot Babbitt metal, making it impossible to remove the irons by key. It of course ruined the leg

Dawson waterfront, May 1900

Seiid-Bassoc Collection, Archives, University of Alaska, Fairbanks

irons but, according to Wiseman, "was well worth the price." Two ex-Mounties who had retired from the force the previous spring were hired to accompany the party the remainder of the way to McNeils Island.

The following morning the party left Dawson aboard the Canadian steamer *Selkirk,* the fastest boat on the upper river. They reached Whitehorse in time to catch the same train for Skagway as the advance party, which had left Dawson two days earlier by a slower boat. The entire party was united again—to the surprise of Marshal Wiseman, who had anticipated a prolonged hunt for Hendrickson and Thornton.

On board the steamer *Jefferson* to Seattle, the most elaborate precautions imaginable were taken for restraint of the prisoners. Hendrickson and Thornton were lodged in a separate stateroom from the other prisoners. The door was removed and a guard was posted where the two prisoners could be watched day and night. Although heavily manacled hand and foot, and despite their hot-Babbitted leg irons, Wiseman made hourly inspections of their chains. On the morning of October 28, without further incident, the *Jefferson* docked at Seattle and the prisoners were transferred to McNeils Island in time for dinner that evening.

One would expect the *Fairbanks Times* to be satisfied with the outcome, but perhaps that is more than should be expected of human nature. Upon learning of Hendrickson's and Thornton's apprehension, the *Times* editorialized, with customary disregard of the facts: "Having at last discovered someone who can catch something besides a bad cold, it is up to Marshal Perry to promptly deputize William Gertz who captured Hendrickson and T. B. Villineve who took Thornton back into custody." The *Times* went on to say that the Hendrickson-Thornton episode demonstrated the futility of crime in the Tanana. With an insight it rarely displayed, it thus bestowed credit where it belonged—on the wilderness, which had probably been the hero throughout.

The agencies of the law were uniformly thankful that the Hendrickson-Thornton chapter was closed, and hopeful that life would resume a more tranquil pattern. Judge Wickersham confided to his diary:

> Have been much disturbed for the last 3 or 4 days. Thornton and Hendrickson, two bad men, sentenced to 15 years each in the U.S. Penitentiary at McNeils Island, Washington, escaped from their guards while on the Str. "Lavelle Young" while being taken out. . . . Read news yesterday that they were recaptured, and they are on their way to McNeils Pen. again. It created much excitement and concern, for it was their fourth escape! Hope they will be landed now. Our office furniture for courthouse has arrived—the last boat of the season has come and gone, and ending the most beautiful October weather I ever saw in the North. We are preparing for winter.

Marshal Wiseman, who probably had more opportunity than any man to know Hendrickson, and most reason to condemn him, gave a most unexpected assessment: "Hendrickson was a good prisoner all of the time he was in my custody. Except in making his attempt to escape at Nation, he did everything he was required to do and caused no unnecessary trouble, the exact opposite from Thornton who was a troublemaker, first, last and all the time. The former was a powerful, dangerous and capable criminal, always on the lookout to attempt an escape, but not a petty one."

EPILOGUE

Fairbanks had not heard the last of the Blue Parka Man's exploits. In December 1906, less than two months after his confinement at McNeils Island, he escaped from custody, but was recaptured before he could leave the island. After that incident Hendrickson was transferred to the Fort Leavenworth, Kansas, Penitentiary, the maximum security prison of its day. That transfer was accomplished with as little delay as possible, on March 18, 1907. Subsequently he made two brief escapes from Fort Leavenworth, the latest on November 13, 1916.

As this news trickled back to Alaska over the years, George Dreibelbis hoped it would not herald a return of Charles Hendrickson. It was not that he feared a renewed outbreak of marauding. After all, there was little left to steal. Gone long ago were the dazzling rich placers, and with it most of the frenzied, boisterous, brawling population of those stampede days. It was said with some if not complete truth that only those remained who did not have the money to leave. Lives sank into narrow, changeless routines. It was no longer a land for a free, restless spirit. Alaska would slumber for another quarter century before it became again the scene of great events.

Dan Sutherland, who had known Charles Hendrickson well in the golden, roisterous days, perhaps said it best. "The old Alaska has passed away. The prospector has departed . . . that in eager, buoyant youth climbed over the mountain passes and on into the vast Interior wilderness beyond, seeking the remote places of the great continent. They were the men who 'fitted in' with their well-spent, free, untrammeled lives."

Thornton died in prison. Hendrickson was released from Leavenworth on February 11, 1920, and was never heard of again.